S0-AUF-338

# The Variation and Adaptive Expression of Antibodies

# The Variation and Adaptive Expression of Antibodies

George P. Smith

Harvard University Press Cambridge, Massachusetts 1973

Library of Congress Catalog Card Number 72–95186
SBN 674-93205-6
Printed in the United States of America

To my mother and father, with love

## Preface

This book treats the phenomenon of adaptive immunity from the viewpoint of molecular genetics. It deals with two basic problems: the origin of antibody diversity, and the selective control of antibody synthesis by antigen. I intend it to be largely self-contained, in that the ideas are developed in the text and the most important primary data are summarized in the tables and figures rather than being merely alluded to. I assume only that the reader is familiar with the fundamental principles of genetics and molecular biology. If he does not know, for example, what a gene (cistron), a codon, the genetic code, a genetic cross, recombination (crossing over), an allele, a genetic map, a chromosome, a polypeptide, messenger and transfer RNA, amino acid activating enzymes, a ribosome, an amino acid sequence, a tryptic peptide, or a disulfide bond is, he will be quickly lost. For those not familiar with elementary immunology, I have included in Chapter 1 a brief introduction to its central features. If this does not prove sufficient, an excellent and relatively brief treatment by Eisen can be found in Chapters 12 to 15 of *Microbiology* by Davis et al. (1967).

I hope too that immunologists who are intimately concerned with the ideas I have dealt with will find in this book an informative account of their field—not because I have introduced anything fundamentally new, but because I have taken pains to bring the many scattered facts and ideas together into a coherent whole, in which the relations of the clues to one another can be appreciated. For despite the aura of paradox that has accompanied the heated debates between countervailing hypotheses in recent years, I believe that we now basically understand the molecular genetics of antibodies at the level we are dealing with; many uncertainties remain, but there is no compelling reason to believe that their resolution will burden us with unsuspected major anomalies.

I wish to thank Hugh McDevitt for my initiation as a votary of theoretical immunology, and Edgar Haber, my thesis advisor, for my training in immunochemistry and protein chemistry. For the last year and a half, as a postdoctoral fellow in his laboratory, I have profited greatly from Oliver Smithies' creativity. In the same period Walter Fitch has taught me the methods of recreating protein evolution, and it was he who supplied the computer programs used in this book. Many of the ideas included here were worked out during fruitful discussions since 1967 with Leroy Hood.

For specific help and advice in putting together this book, I am indebted to William Gray, E. H. Peters, Robert Goodfliesh, Thomas Wegmann, David Zoschke, and David Gibson. Karen Hamerslag did the typing, my father drew the original figures, Steven Hong drew the final figures, and E.H. Peters composed the index.

Finally, I thank my countrymen, who through their proxy, the National Institutes of Health, have seen fit to support me as a graduate student and postdoctoral fellow, and the Helen Hay Whitney Foundation for a generous postdoctoral fellowship. The University of Wisconsin Computing Center, whose facilities were used, is supported in part by grants from the National Institutes of Health and the National Science Foundation. This book is publication number 1489 from the Laboratory of Genetics of the University of Wisconsin.

## Contents

## Tables and Figures

Appendix Tables and Figures

# The Variation and Adaptive Expression of Antibodies

# Chapter 1

## Introduction to the Immune Response

When a foreign substance, called an antigen, is injected into an immunologically competent animal, the animal responds by the synthesis of proteins called antibodies, which are able to bind the antigen. When the same antigen is injected again some time later, the synthesis of antibodies typically starts sooner, reaches higher levels, and is more sustained than in the primary response (Fig. 1-1). This so-called secondary response implies that the animal "remembers" in some way his previous encounter with the antigen, and it has therefore been called immunological memory.

Antibodies belong to a family of related proteins, which are called immunoglobulins. The term "antibody" has in the past been reserved for the immunoglobulin formed in response to an antigen, to distinguish it from immunoglobulin for which no antigen is known. As there is no evidence that there is anything essentially different about the latter, however, I shall use the terms "antibody" and "immunoglobulin" interchangeably.

The antibody response is highly specific for the injected antigen. The antibody induced by one antigen only rarely binds any other antigen, and these infrequent so-called cross-reactions have almost always been attributable to obvious structural similarities between the cross-reacting antigens. Immunological memory, too, is highly specific. If the animal is "primed" by a first injection of antigen $A$, he will mount a typical secondary response to a second dose of antigen $A$, but a typical primary response to antigen $B$.

The immune apparatus is xenophobic: an animal is not ordinarily stimulated to form antibodies against his own substances, even though these may be highly antigenic in another animal. The ability to distinguish between foreign and endogenous antigens is called immunological tolerance, and without it the immune response would wreak havoc on the animal. Tolerance, like the antibody response and memory, is highly specific: an animal tolerant to his own antigens may mount a vigorous response to foreign antigens quite closely related to his own. Cases are

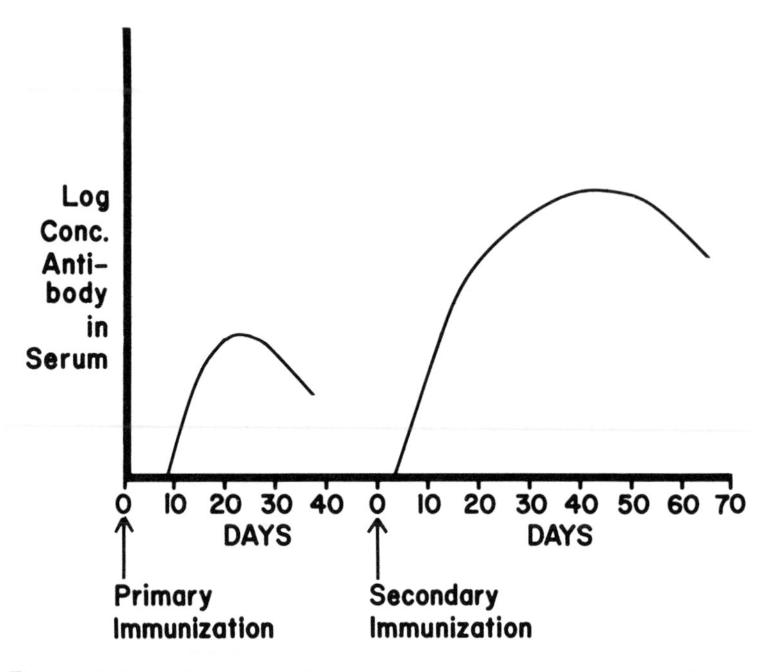


**Figure 1-1.** Schematic diagram of primary and secondary (anamnestic) antibody responses. The day numbers are intended to give only a rough idea of the times involved. The absolute values of the antibody levels are highly variable; the peak level seldom exceeds 1 or 2 mg antibody per ml of serum, except after special hyperimmunization procedures.

known where an animal may respond to a protein different from his own by only a single amino acid replacement.

Tolerance is in reality a state of negative immunity, for it arises only in response to exposure of the animal's immune apparatus to the tolerogenic antigen. Thus an animal is in general not tolerant to endogenous antigens which are sequestered from its immunological apparatus, but is tolerant to endogenous antigens which are not. Animals may be made experimentally tolerant to foreign antigens if they are administered in certain ways; for instance, antigens injected early in development and able to persist in the body readily elicit long-lasting tolerance. Immunity and tolerance thus are two alternative responses to an antigen: what exactly determines which of these two paths the immune apparatus will take is an outstanding problem in immunology.

All immune responses share these three signal characteristics—specificity for the antigen, memory, and tolerance—but they may assume a wide variety of forms. Antibodies against bacteria or other cells may effect their lysis by a series of blood enzymes collectively called complement. Antibodies may combine with toxins, neutralizing their deleterious effects. They may precipitate or agglutinate macromolecular or particulate antigens (respectively), forming a large complex which can be rapidly engulfed by phagocytic cells. They may, even without agglutination, facilitate phagocytosis of certain harmful bacteria such as pneumococci. A special "reaginic" class of antibodies to various antigens is responsible for allergies such as hay fever. There are indeed certain immune responses, recognized as such by their specificity, memory, and tolerance, which are not attributable to *humoral* antibodies (that is, antibody molecules dissolved in the body fluids) at all, but rather are associated with cells. Most immunologists agree that cell-bound immunity involves antibodies which are tightly bound to the surface of the immune cells and are not secreted in a soluble form. "Delayed" hypersensitivity to some antigens such as tuberculin, which superficially resembles antibody-mediated skin inflammation, is an example of cell-bound immunity; the rejection of transplanted foreign tissues and organs is also thought to be largely mediated by immune cells rather than soluble antibodies.

In each of these immune phenomena the primary event, which determines the specificity of the reaction and from which the other diverse manifestations of immunity follow, is the combination of the antibody with the antigen. Like all cases of biological specificity, the ability of an antibody to discriminate between the homologous antigen and other substances results from the complementary nature of the two reacting surfaces: in Ehrlich's (1900) words, "the groups . . . must be adapted to one another, e.g. . . . as lock and key." The work of Karl Landsteiner (1945) between the First and Second World Wars has led to a more sophisticated analysis of the specificity of antibody-antigen reactions. Landsteiner exploited the fact that small organic molecules, called *haptens*, when coupled to macromolecular *carrier* molecules, can elicit the formation of antibodies specific for the hapten moiety alone. The experimental fact of hapten-specific antibodies has been generalized to all antibody-antigen reactions. Individual antibody molecules are not directed to a whole complex antigen molecule, but rather to small areas on its surface, called determinants. A complex antigen therefore will elicit a large variety of antibodies, directed against its myriad surface determinants. The hypothesis of antigenic determinants has been amply vindicated in the ensuing years.

In this book, in which the phenomenon of immunity is considered almost exclusively from the standpoint of molecular genetics, the cells responsible for the immune response appear as abstract entities best characterized as bags of genes. Nevertheless, our knowledge of them is more substantial. Immunity is a property of the lymphoid system, which consists of the lymphoid organs (spleen, lymph nodes, appendix, thymus, bone marrow, the bursa of Fabricius in birds, and some other organs) and the lymphoid cells (lymphocytes) circulating in the blood and lymphatics. These organs are further subdivided into the *peripheral* lymphoid system (primarily spleen, lymph nodes, and circulating lymphocytes), which actually mount immune responses, and the *central* lymphoid system (primarily thymus, bone marrow, and the bursa in birds) which is necessary for the proper development of immunocompetence. Lymphocytes appear to derive from the bone marrow and thymus throughout the life of the organism (perhaps also from other organs such as fetal liver early in ontogeny), and to turn over extremely rapidly, with rapid cell division and frequent cell death more or less balancing each other. Lymphoid cells, whether in the organs or in circulation, form a continuous spectrum of cell types ranging from small lymphocytes (little cytoplasm, low concentration of protein synthesizing machinery) to large lymphocytes (more cytoplasm and synthetic apparatus) to *plasma cells* (very high concentration of protein synthesizing machinery, complex endoplasmic reticulum), which are responsible for the bulk of antibody synthesis. Lymphoid cells are also divided into the functional classes *B* and *T* on the basis of inferences from various experiments (Chapter 8). Apart from the lymphocytes, macrophages are also thought to play a key role in immunity. These are large phagocytic cells, found in both sessile and mobile states. They are thought by many to present engulphed antigens in a highly immunogenic form to lymphoid cells; in addition, a more fundamental role in the immune response is sometimes imputed to them on the basis of some very surprising results which will be discussed in Chapter 8.

## Preadaptive Immunocompetence and the Central Problem of Immunity

The immune apparatus responds adaptively to almost any arbitrarily chosen antigen, even artificial haptens. With the learning response, then, it shares the ability to "expect the unexpected," for it is most unlikely that any but a few of the antigens to which an animal can respond actually played any significant selective role in its evolutionary past (Cohn, 1968).

Borrowing a term from bacterial genetics, I shall call this property of immunity "preadaptive" competence; for the ability to respond to any particular antigen evidently arises prior to any evolutionary need to do so, in the same way that bacterial mutations arise preadaptively, independent of their survival value.

I shall follow the current approach to the problem of adaptive immunity by dividing it into two parts, the diversification and selection of biological information (Fig. 1-2). On the one hand, we must account for the diversification of information leading to the various antibodies with their distinct specificities. On the other, we must explain how the expression of this information is specifically controlled by antigen, which in some way must induce the expression of that information which leads to the appropriate antibodies. In order to account for the tolerance of an immunocompetent animal to his own "self-antigens," there must also be another aspect of antigenic control, by which the expression of inappropriate information is suppressed.

The problem of adaptive immunity thus has two aspects, the problem of antibody diversity and the problem of selection by antigen. For both of these, the preadaptive nature of immunocompetence poses special difficulties. Preadaptive competence implies that the mechanism by which any particular antigen stimulates the synthesis of only the appropriate antibodies must in general have arisen before it conferred any survival value; this fact imposes constraints on any theory of antigenic selection, which

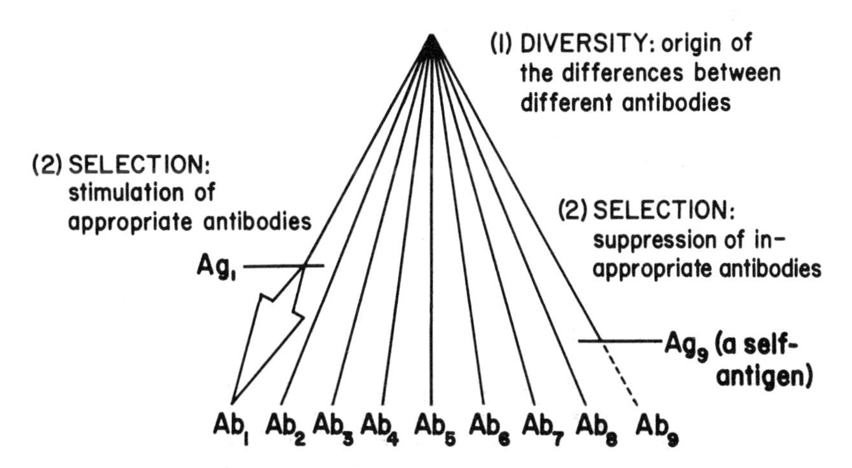


**Figure 1-2.** Two central problems of adaptive immunity: that of antibody diversity and that of selection by antigen.

(as I shall argue in Chapter 8) are elegantly met by the clonal selection theory. This hypothesis is now almost universally accepted by immunologists, and the bulk of the present book will be concerned with the more controversial problem of antibody diversity.

## The Problem of Antibody Diversity

> The unlikelihood of any direct action of the antigen in the organism led Ehrlich to the conclusion that small quantities of antibodies are preformed in the animal, and that the injection of antigen merely increases the level of normal antibody-synthesis . . . This hypothesis could be accepted as long as only a few natural antigens were known; since the work of Landsteiner and his coworkers, however, an almost unlimited number of artificial antigens . . . can be made which have never appeared in nature. It is not reasonable that an animal produces predetermined antibodies against thousands of such synthetic substances.
>
> —Breinl and Haurowitz (1930); my translation

The basic theorem of evolution—all life from pre-existing life—has its molecular counterpart: all DNA from pre-existing DNA. It is assumed, and there is much evidence for this view, that all biological information has descended from a small amount of ancestral information by an ordered series of twofold divergences, each of which represents in principle the replication of a molecule of DNA. Thus the provenance of contemporary DNA (and the RNA and protein molecules copied from it) can be represented as a genealogy or family tree with multiple branches: what Fitch and Margoliash (1968) have called the actual family tree. The diversity of this information arises in small, discrete steps (mutations), each of which is suffered by a particular molecule at a particular time and is thus ascribable to a particular branch on the actual family tree. The endless proliferation of information is more or less counterbalanced by the constant death of certain branches, because of the death of organisms and the ensuing degradation of their DNA.

In multicellular higher organisms, most of the divergences in the actual family tree correspond to cell divisions during the life of individual organisms. Only a few of the branches lead to the germ cells; they will be collectively called the *germline*. Mutations occurring in germline branches can be inherited by the descendants of the individuals in which they occur. Mutations occurring in other branches, which will be called *somatic*, are only transmitted to some of the cells of the particular individuals in which they arise.

It would of course be impossible to reconstruct the actual family tree. However, as we shall see, it is possible to use our knowledge of a limited number of contemporary genes to infer the connection of their lineages back through the complexities of the actual family tree.

Antibodies form a family of related proteins collectively called immunoglobulins: their strong homology to one another leaves no doubt that they descended from a common ancestor (see Chapter 2). The problem of antibody diversity reduces to the question of where and how in the actual antibody family tree their diversity is introduced. The germline hypothesis of diversity supposes that it arises in the germline and is propagated or discarded by natural selection depending upon whether it enhances or detracts from the overall fitness of those individuals in which it is present. According to this hypothesis, genetic information for the entire range of antibodies that an individual will be able to make is already encoded in the DNA of the zygote, which must therefore contain diverse structural genes for the various antibodies. This is accomplished by the process of gene duplication, diagrammed in Fig. 1-3. Hence the germline hypothesis is sometimes called the multiple gene hypothesis.

In a sense, the germline theory must be regarded as the conservative model, for it conforms to our traditional notion of how biological diversity arises. It implicitly formed part of Ehrlich's (1900) theory of antibody formation as early as the turn of the century. By the time the preadaptive nature of immunocompetence became evident, however, the germline

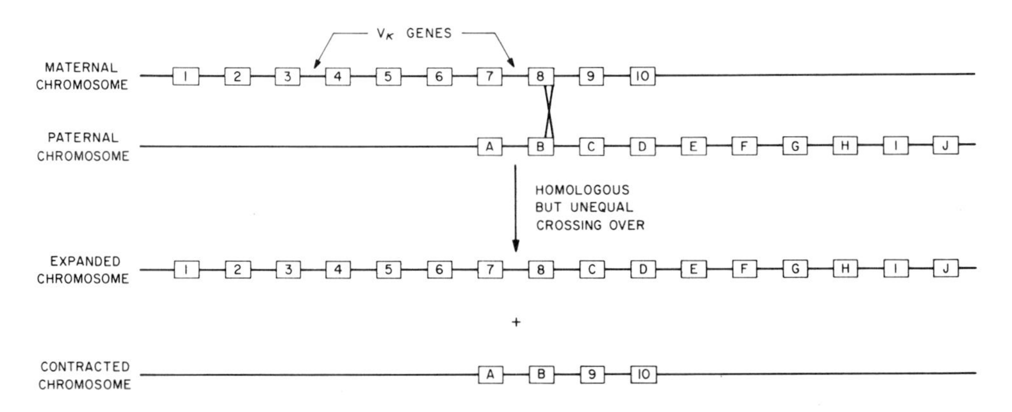


**Figure 1-3.** Gene duplication and deletion. Unspecified chromosomal rearrangements can lead to chromosomes with two adjacent copies of some gene. Once this state is reached, out-of-register crossover between homologous genes can lead to very rapid shifts in the number of genes per chromosome, both because homologous crossover is a relatively frequent event and because each such event can duplicate or delete a whole block of adjacent genes. Taken from Smith, Hood, and Fitch (1971).

hypothesis was singled out from among the implicit assumptions of the older immunologists, and called into serious question. For example, Breinl and Haurowitz (1930), in the passage quoted at the beginning of this section, argued essentially that preadaptive immunocompetence vitiates any germline hypothesis because it implies (*a*) that antibody diversity is limitless and therefore not encoded in the necessarily limited germline genome, and (*b*) that antibody diversity arises preadaptively and therefore cannot be attributed to natural selection acting on the level of the whole organism.

In the light of our more sophisticated understanding of the immune response, neither of these arguments is compelling, as will be discussed in Chapter 5. Nevertheless, Breinl and Haurowitz's paper must be regarded as a signal event in the history of theoretical immunology, for it ushered in forty years of speculation about alternative theories of diversity, which would account for its apparent limitlessness, and which would not require that conventional natural selection maintain germline diversity. Collectively these theories are called *somatic* hypotheses, since they all suppose that antibody diversity arises separately in each individual.

The first somatic theory was the template hypothesis of Breinl and Haurowitz themselves (1930), later supported by Pauling (1940). In this model antibody chains as originally synthesized are capable of assuming various final three-dimensional forms. These unformed chains fold around an invading antigen into a stable complementary configuration and thus assume their final structure and specificity for the antigen. According to the template model, therefore, antibody diversity arises in the multifarious three-dimensional configurations assumed by a single polypeptide chain under the influence of various antigens.

The template theory is no longer tenable, for it has been shown that the specificity of antibody is determined by its amino acid sequence. Haber (1964) and Whitney and Tanford (1965) completely unfolded antibodies in strong denaturing solutions with cleavage of all the disulfide bonds, so that all structural information other than the primary amino acid sequence was lost; yet these antibodies regained their specificity on being returned to a physiological medium in the absence of antigen. This result has since been repeated several times (Freedman and Sela, 1966; Jaton et al., 1968). Moreover, there is considerable diversity of immunoglobulin amino acid sequence, and careful amino acid analyses by Koshland and her colleagues have uncovered compositional differences (thus sequence differences) which correlate with the specificity of the antibody (Koshland, 1967; Koshland, Davis, and Fugita, 1969; Koshland and Englberger, 1963; Koshland, Englberger, and Shapanka, 1966; Koshland, Reisfeld, and Dray, 1968). Hence, contrary to the template hypothesis, the diversity of antibody specificity reflects a corresponding diversity of antibody amino acid sequences, and modern somatic theories must explain how this diversity arises. The most important of the newer hypotheses are the *somatic mutation* theory proposed by Burnet (1959) as part of the clonal selection theory, and the *somatic recombination* hypothesis, proposed first by Smithies (1963). A major portion of this book will be devoted to an attempt to choose between the germline theory and these somatic alternatives.

### The Number of Antibodies

In order to put the problem in perspective, let us discuss the number of different antibodies which must be accounted for. Clearly this number must be large to explain the extraordinary flexibility of the immune response. However, the following estimate of the magnitude of antibody diversity, which is based instead on the variability of immunoglobulin amino acid sequences, far exceeds any number necessitated by the facts of serological specificity, which can quite plausibly be accounted for by a "combinatorial" scheme requiring a relatively limited number of antibodies (Chapter 5).

As we shall see in the next chapter, the amino acid sequence variation which might account for the diversity of antibody specificities is confined to certain regions of antibody molecules (the V regions) which fall into three major families: $V_\kappa$, $V_\lambda$, and $V_H$. Taking the first two of these, and confining our attention to humans, we may set a lower limit on the total number of human $V_\kappa$ and $V_\lambda$ sequences by insisting that it be large enough to account for the fact that no two randomly selected chains have yet been demonstrated to be identical. From these considerations the total number of human $V_\kappa$ and $V_\lambda$ sequences must almost certainly be larger than 425 to explain the results of Quattrocchi, Cioli, and Baglioni (1969), who demonstrated that all 52 human $V_\kappa$ and 50 human $V_\lambda$ sequences studied by them are different. Theories of antibody diversity, then, must account for about 1,000 different V regions in humans. An antibody molecule has two different V regions (one $V_\kappa$ or $V_\lambda$, and one $V_H$), and if we suppose that both can influence the binding specificity, the number of different antibody combining sites may greatly exceed the total number of V regions.

This may seem like an unacceptably large number if each V region is coded by its own germline gene. But we must keep these numbers in perspective. If the collective patience of immunologists were sufficient to determine 600 independent human $V_\kappa$ sequences, and each were shown to be different from all the others, the total number of inferrable $V_\kappa$ regions ($6 \times 10^4$; see Fig. 1-4) would still account for only about 12 percent of an average human chromosome.

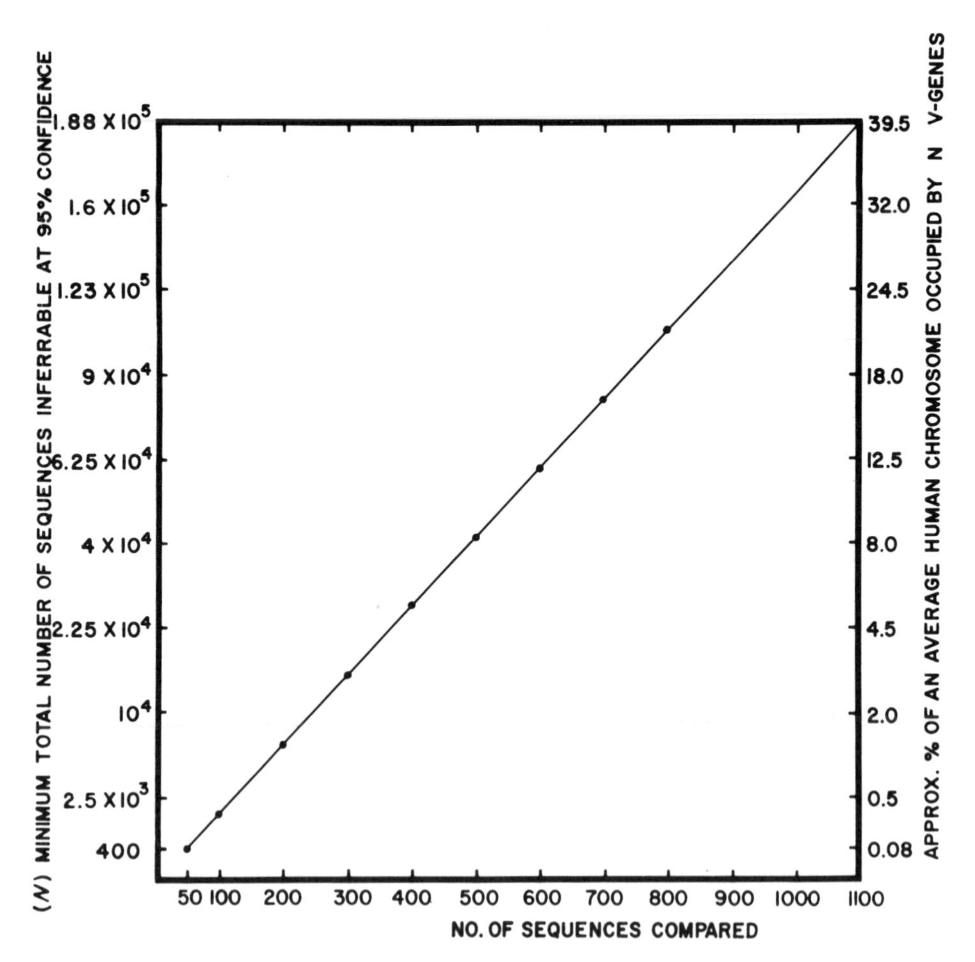


**Figure 1-4.** Number of sequences (*N*) inferrable at 95-percent confidence level when various numbers of sequences are compared and found nonidentical.

Nevertheless, 1,000 different V regions is a large amount of amino acid sequence variation. After outlining the general structure of antibody molecules, methods of reconstructing their evolutionary history, and the evolution of C regions, we shall turn to the details of the theories which have been proposed to account for it.

## Chapter 2

# The Basic Structure of Immunoglobulins

All immunoglobulins are composed of one or more monomers with the general structure shown schematically in Fig. 2-1. Each monomer consists of four polypeptide chains, two identical light (L) and two indentical heavy (H) chains. Both types of polypeptide chain appear to be composed of linearly connected units consisting of about 110 residues each: L chains contain two such units (about 22,500 daltons), while $\gamma$-type H chains contain four such units (carbohydrate-free molecular weight, about 47,000 daltons). The number of units in other types of H chain has not been determined.

The four polypeptides are arranged into three physical domains as shown in Fig. 2-1. Each of the two Fab domains contains one antigen-binding site, and the Fc domain seems to be important for many of the "effector functions" (Edelman et al., 1969): that is, functions which antibodies carry out regardless of specificity. This structure was originally based on the observation of Porter (1959) and others that several proteolytic enzymes consistently liberated from the native molecule intact globular fragments corresponding to the three domains, indicating preferential cleavage in a presumably extended portion of the H chain designated as the "hinge." The sites of preferential cleavage of several native immunoglobulin molecules by several enzymes are shown in Fig. 2-2. The three-domain structure is also consistent with recent electron microscopic and x-ray crystallographic data (Fig. 2-1).

The four-chain/three-domain structure must be an ancient property of the immunoglobulins. It is found in all known classes of immunoglobulins in mammals (IgG, IgA, IgM, IgE, and IgD) and in vertebrate species ranging from man to the lamprey (Table 2-6). The quaternary structures of mammalian IgG, IgM, and IgA differ greatly, however, as shown in Fig. 2-3.

One prediction of this unitary view is that IgM molecules–which are pentameric and therefore contain ten Fab domains–should show ten binding sites, an expectation which has sometimes been confirmed and sometimes not; the meaning of the reported cases of less than ten binding sites per molecule is not yet clear and is discussed in a recent review by Metzger (1970).

In any one immunoglobulin molecule all the H chains and all the L chains

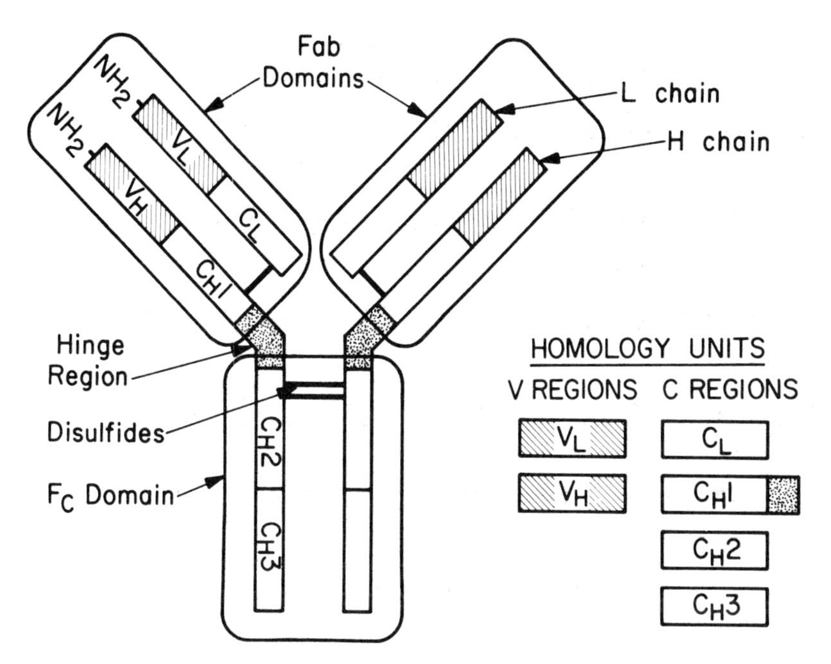


**Figure 2-1.** The basic four-chain/three-domain immunoglobulin monomer. Evidence for this structure, reviewed by Dorrington and Tanford (1970), comes from Fig. 2-2, and from x-ray crystallographic (Avey et al., 1968; Goldstein, Humphrey, and Poljak, 1968; Humphrey, 1967; Humphrey et al., 1969; Terry, Matthews, and Davies, 1968) and electron microscopic (Chesebro, Bloth, and Svehag, 1968; Feinstein and Munn, 1969; Green, 1969; Parkhouse, Askonas, and Dourmashkin, 1970; Shelton and McIntire, 1970; Svehag and Bloth, 1970; Svehag, Bloth, and Seligmann, 1969; Svehag, Chesebro, and Bloth, 1967; Valentine and Green, 1967) studies. Taken from Smith, Hood, and Fitch (1971).

are identical. This is perhaps best shown by the finding of only a single L chain and a single H chain sequence in myeloma proteins. For example, the entire sequence of one myeloma protein has been determined (protein Eu; Edelman et al., 1969) and no evidence for two different L or H chains has been found. The identity of the two H chains and two L chains implies a twofold rotational symmetry for both the whole molecule and the Fc domain, and this expectation has been confirmed by x-ray crystallographic analysis (references in Fig. 2-1).

Normal immunoglobulin molecules probably comprise two identical halves. For example, Gilman, Nisonoff, and Dray (1964) showed that individual immunoglobulin molecules from heterozygous rabbits are symmetric with respect to two allelic forms of polypeptide chains; and Fudenberg, Drews and Nisonoff (1964) showed that the binding sites of individual anti-

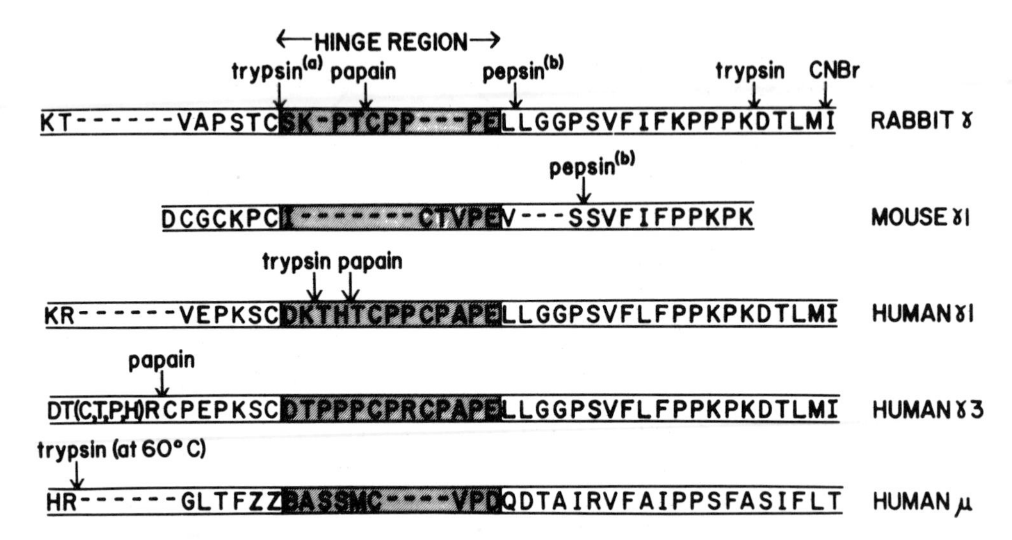


**Figure 2-2.** Sites of preferential hydrolysis of several native immunoglobulins. Taken from Edelman et al. (1969); Frangione and Milstein (1969); Givol and DeLorenzo (1968); Press and Hogg (1969); Shimizu et al. (1971); Svasti and Milstein (1970); Turner and Bennich (1968); Turner et al. (1969); Utsumi (1969).

[a]When Cys is aminoethylated.

[b]Pepsin also cleaves within the Fc domain, liberating small fragments and one large fragment (pFc′) corresponding to a dimer of $C_H3$.

bodies in an animal responding to two different types of red blood cells are directed against one or the other of the red cells, but not both. The specialization of individual antibody-producing cells to a single antibody specificity, a single class of H and L chain, and (in heterozygotes) a single allelic form also implies the symmetry of normal immunoglobulin molecules (Chapter 8).

Immunoglobulin polypeptide chains can be divided into *homology units* on the basis of evolutionary homology (Edelman et al., 1969), as shown in Fig. 2-1. (Two sequences are presumed to derive from a common ancestor if, by the criterion of Fitch, 1966, they are significantly more similar in sequence than would be expected by chance from their respective amino acid compositions.) Each unit is about 110 residues long, and it is presumed that all immunoglobulins are constructed of them. The N-terminal units of both H and L chains are homologous to one another and are called V regions ($V_H$ and $V_L$ respectively), while all the remaining units are likewise homologous to one another and constitute the C regions. Thus the C region of an L chain ($C_L$) consists of one homology unit, while that of

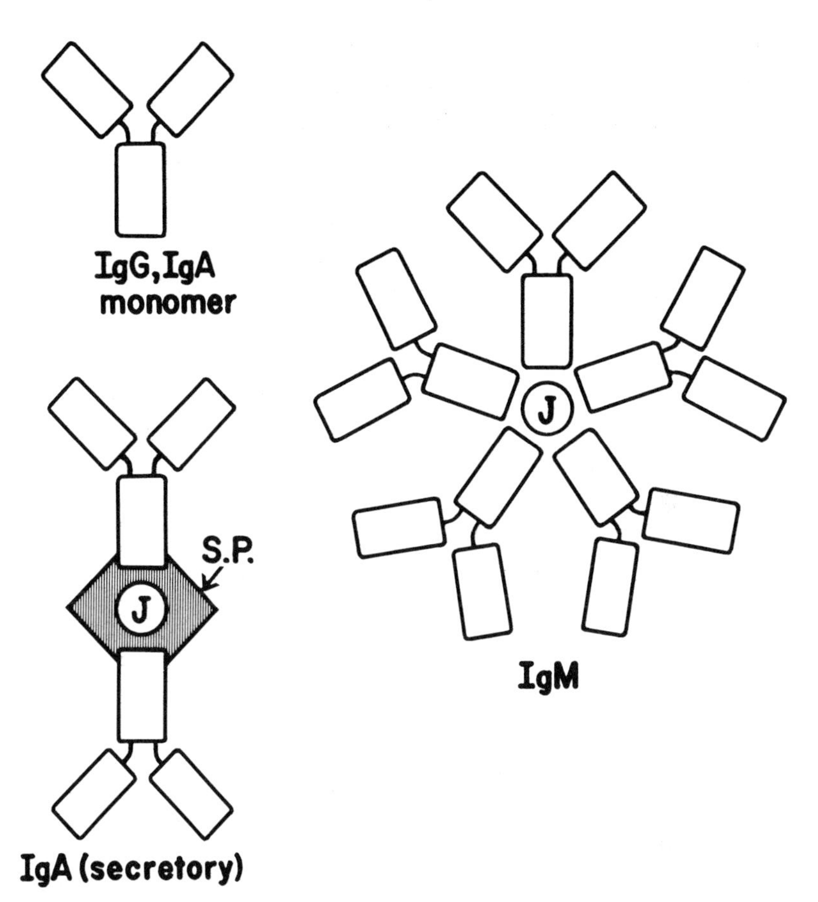


**Figure 2-3.** Quaternary structure of three classes of immunoglobulin. *J* = J chain (Halpern and Koshland, 1970; Mestecky, Zikan, and Butler, 1971), *S.P.* = secretory piece. Evidence for these structures is cited in Table 2-6; they are also supported by electron microscopic studies (references cited in Fig. 2-1).

an H chain ($C_H$) contains three or four ($C_H1$, $C_H2$, $C_H3$, and $C_H4$). These homologies imply that V regions evolved from a common ancestral V gene, and likewise that C regions evolved from a common pristine C gene.

There is no certain statistical evidence for evolutionary homology between contemporary V and C regions (Edelman et al., 1969). However, the very similar lengths of the two types, and their common pattern of intrachain disulfide bonds (Fig. 2-4), strongly suggest that both derive from a single primordial immunoglobulin gene, as first suggested by Hill and his associates (1966). If so, many subsequent mutations in V and C regions have obscured the original sequence homology between them.

There is a short segment of $C_H$, more or less coextensive with the region of preferential hydrolysis of the native molecule (Fig. 2-2), which has no apparent homology with any other portion of the molecule (Edelman et al., 1969; see Fig. 2-1); this segment will be called the "hinge" and will be considered part of $C_H1$.

The classification of immunoglobulins has undergone considerable change as our knowledge of them has increased: consequently, a number of overlapping taxonomies are still current. Table 2-1 summarizes the nomenclature I shall use.

A growing body of evidence points to three major genetically functional units, which I shall call gene families: one such family controls the H chains and will be called the H family; and two, the $\kappa$ family and the $\lambda$ family, control the L chains. Tables 2-2, 2-3, and 2-4 show most of the available amino acid sequences for $\kappa$, $\lambda$, and H chains respectively. As for all amino acid sequences in this book, these are given in the single-letter amino acid code recommended by the International Union of Biochemists and indicated in the legend to Fig. 3-2. Appendix Tables A-1, A-2, and A-3 list these proteins, their properties and provenances.

Each immunoglobulin family is characterized by a distinctive family of V regions. These families, $V_\kappa$, $V_\lambda$, and $V_H$, can be further subdivided into several *subgroups* of V regions, which are closely related in sequence, as will be discussed in Chapter 6. These subgroups are designated by roman numeral subscripts, such as $V_{\kappa I}$, $V_{\kappa II}$, and $V_{\kappa III}$.

The three immunoglobulin families are also characterized by a distinctive family of one or more *classes* of C region. The $\kappa$ and $\lambda$ families contain one class each, $C_\kappa$ and $C_\lambda$, while the H family contains several classes, $C_\gamma$, $C_\mu$, $C_\alpha$, $C_\delta$, $C_\epsilon$, and perhaps some as yet unrecognized ones.

The major C-region classes in several mammalian species can be divided into subclasses, which are listed in Table 2-5. Each of these subclasses is found in all individuals of the species, and therefore they are nonallelic gene duplications. In general, subclasses of distantly related mammals do not correspond to one another, so that the responsible gene duplications are quite recent (Chapter 4). These subclasses will be designated by subscripts, such as $C_{\lambda arg}$, $C_{\gamma 1}$, $C_{\gamma 2}$, $C_{\gamma 3}$.

An immunoglobulin polypeptide chain, which consists of one V region and one C region from the same gene family, is named after its C region: thus $\kappa$ chains have a $C_\kappa$ region, $\gamma_4$ chains have a $C_{\gamma 4}$ region, and so on.

Finally, a whole immunoglobulin molecule, which consists of two identical L chains (either $\kappa$ or $\lambda$) and two identical H chains, is classified

**Table 2-1.** Nomenclature for immunoglobulins.

| Gene family | V-region family | Subgroups | C-region classes | C-region subclasses | Polypeptide chains | Immunoglobulin classes |
|---|---|---|---|---|---|---|
| H | $V_H$ | $V_{HI}, V_{HII}, V_{HIII}$ | $C_\gamma$ | $C_{\gamma 1}, C_{\gamma 2}, C_{\gamma 3}, C_{\gamma 4}$ | $\gamma_1, \gamma_2, \gamma_3, \gamma_4$ | $IgG_1, IgG_2, \ldots$ |
| | | | $C_\mu$ | | $\mu$ | IgM |
| | | | $C_\alpha$ | $C_{\alpha 1}, C_{\alpha 2}$ | $\alpha_1, \alpha_2$ | $IgA_1, IgA_2$ |
| | | | $C_\delta$ | | $\delta$ | IgD |
| | | | $C_\epsilon$ | | $\epsilon$ | IgE |
| $\kappa$ | $V_\kappa$ | $V_{\kappa I}, V_{\kappa II}, V_{\kappa III}$ | $C_\kappa$ | | $\kappa$ | Found |
| $\lambda$ | $V_\lambda$ | $V_{\lambda III}, V_{\lambda V}, \ldots$ | $C_\lambda$ | $C_{\lambda arg}, C_{\lambda lys}$ | $\lambda_{arg}, \lambda_{lys}$ | in all classes |

Gene family: Set of multiple linked genes, whose V and C genes are expressed colinearly in the same polypeptide chain.

Subgroups: Sets of closely related V regions within one family. Named by roman numeral subscript.

Subclasses: Minor, nonallelic variants of several of the major C-region classes, caused by relatively recent gene duplications. Named by subscript (not roman numeral).

Homology units: Segments of immunoglobulin chains about 110 amino acids long, which derive from a progenitor gene common to all immunoglobulins. L chains contain two: $V_L$ and $C_L$. $\gamma$ chains contain four: $V_H$, $C_H1$, $C_H2$, and $C_H3$.

Position numbers: Unless otherwise indicated, refer to Table 2-2 ($\kappa$ chains, numbering system for protein Ag), Table 2-3 ($\lambda$ chains, numbering system of protein Ha), and Table 2-4 (H chains, numbering system of protein Eu).

| PROTEIN | SUB GRP | ALLO-TYPE | 1 2 3 3 <br> 0 1 2 3 4 5 6 7 8 9 0 1 2 3 4 5 6 7 8 9 0 1 2 3 4 5 6 7 8 9 0 1 a b c d e f 2 |
|---|---|---|---|
| Joh | I | | D I Q M T Q S P S S L S A S V G D R V T I S C |
| Lay | I | | D I Q M T Q S P S S L S V S V G D |
| Pap | I | | D I Q M T Q S P S S L S V S V G D R V T I A |
| Hau | I | Val | D I Q M T Q S P S S L S A S V G D R V T I T C R A S Q S I S S - - - - - - Y |
| Ou | I | Val | D I Q M T Q S P S S L S A S V G D R V T I T C R A S Z T I S S - - - - - - Y |
| Ale | I | | B I Z M T Z S P S S L S A S V G D R/V T L T C R |
| Dav | I | | D I Q M T Q S P S S L S T V V G D R V T I T C D A S Q B I B S - - - - - - W |
| Fin | I | | D I Q M T Q S P S S L S T V V G D R V T I T C D A S Q B I B S - - - - - - W |
| Car(1) | I | | B I Z M T Z(S,P,S)T L S A S V G D R/V A I T C R |
| Eu | I | Val | D I Q M T Q S P S T L S A S V G D R V T I T C R A S Q S I N T - - - - - - W |
| Paul | I | | D I Q M T Q S P S T L S A S V G D R V T I T C R A S Q S I S S - - - - - - S |
| HBJ 4 | I | Leu | D I Q M T Q S P S T L S A S V G D R V S I A C R A S Q B V S N - - - - - - W |
| Mon | I | | D I Q M T Q S P S T L S A S V G D R V T I T |
| Mar | I | | D I Q L T Q S P S S L(S,A,S,V,G,D) |
| Bel | I | Val | B I Z L T Z S P S S L S A S V G D R/V T I T C Z A S Z B I S K |
| Lux | I | | D I Q L T Q S P S F L S A S V G D R V T I T |
| Wag | I | | D I Q M T Q S P S S L S A S V G D |
| HBJ 10 | I | | D I Q M T Q S P S S L S A S(V,G,B)R |
| Au | I | Val | D I Q M T Q S P S S L S A S V G D R V T I T C Q A S Q D I S D - - - - - - Y |
| Ag | I | Val | D I Q M T Q S P S S L S A S V G D R V T I T C Q A S Q D I N H - - - - - - Y |
| Roy | I | Leu | D I Q M T Q S P S S L S A S V G D R V T I T C Q A S Q D(I,S)I - - - - - - F |
| Pot | I | | D I Q M T Q S P A S L S A S V G D R V T I T C |
| Con | I | | D I Q M T Q S P S S L S A S V G D R V T I T |
| Tra | I | | D I Q M T Q S P S S L S A S V G D R V T I T |
| HBJ 1 | I | | D I L M T Q S P T S L S A S V G D R |
| Car(2) | I | | D I Q M T Q S P S S L S A S V G D R V T I T C |
| Tel | I | | D I Q M T Q S P S S L S A S V G D R V T I T C |
| Die | I | | D I Q M T Q S P S S L S A S V G D R V T I T C |
| Cra | I | | D I Q M T Q S P S S L S A S L R D R V T I T |
| Ker | I | Leu | B(I,Z,M,T,Q,S,P,S,S,L)S A S V G D R I T I T C Q A S Q D I K(N,- - - - - - F) |
| BJ | I | Leu | B(V,Z)M T Q(S,P,S,S,L,S,A,S,V,G,D)R V T I T C Q A S Q D I N K - - - - - - Y |
| Dee | I | | B I Z M T(Z,S,P,S,S,L)S A S V G D R/V T I T C R |
| Ti | II | Val | E I V L T Q S P G T L S L S P G E R A T L S C R A S Q S V - - S N S - - - F |
| Fr 4 | II | Leu | E(I,V,L)T Q S P G T L S L S P G E R/A T L S C R/A S Q S V - - R N N - - - Y |
| Ste | II | Val | E I V L T Q S P G T L S L S P G E R A A L S |
| B6 | II | Val | (Z,I,V,L,T,Z,S,P,G,T,L,S,L,S,P,G,Z,R/A A L S C R A S Q S L - - S G N - - - Y |
| Dil | II | Val | E I V L T Q S P G T L S L S P G D R A T L S C R A S Q S L - - B S K - - - S |
| Rad | II | Val | E I V L T Q S P G T L S L S P G D R A T L S C R A S Q V S - - S N S - - - Y |
| Cas | II | | E I V L T Q S P G T L S L S P G D R A T L S |
| HBJ 5 | II | | E I V L T Q S P B T L S L S P G E R |
| HS 4 | II | | E I V L T Q S P G T L S L S P G E R |
| How | II | | E I V L T Q S P G T L S L S P G E |
| Nig | II | | K I V L T Q S P G T L S L S P G E R A T L S |
| Hac | II | Val | E I V L T Q S P G T L A L A P G E R A |
| Win | II | | D I V L T Q S P A T L S L S P G E R A T L S |
| Smi | II | | E I V L T Q S P A T L S L S P G E R A T L S |
| Gra | II | | E M V M T Q S P A T L S M S P G E R A T L S |
| Dob | II | | E I I M T Q S P A |
| Gal | III | Leu | Z I V M T(Z,S,P,L)S L P V(T,P,G,Z,P,A,S,I,Z,C)R |
| Mil | III | Val | D I V L T Q S P L S L P V T P G E P A S I S C R/S S Q N L L/Z S B G B - - Y |
| Tew | III | Val | D I V M T Q S P L S L P V T P G E P A S I S C R S S Q - - H(G,B)S - - - - F |
| Rai | III | Leu | B I V M T Z S P L S L P V T P G E P A S I S C R |
| Man | III | Val | D(I,V,M,T,Q,S,P,L)S L P V T P(G,E,P,A,S)I S C R |
| Bat | III | | D I V M T Q S P L S L P V T P G E P A S I S C R S S Q(S)L L H(S)B G B B - Y |
| Cum | III | Val | E D I V M T Q T P L S L P V T P G E P A S I S C R S S Q S L L D S G D G N T Y |
| MOUSE | | | |
| MBJ 41 | I | | D I Q M T Q S P S S L S A S L G E R V S L T C R A S Q B I G S - - - - - - L |
| MOPC173 | I | | D I Z M T Q T T S S L S A S L G D ? V T I |
| MOPC31C | I | | D I Q M T Q S P A S L S A S V G E R V T I T C ? A |
| MOPC23 | I | | D I Q M T Q S P A S L(S)V(S)V G Z ? V T I |
| MOPC149 | I | | D I Q M T Q S P B Y L(S)A(S)V ? E R V T I(T,C,R) |
| MBJ 70 | II | | D I V L T Q S P A S L A V S L G Q R/A T I S C R/A S E S V B B S G I S - - F |
| MOPC321 | II | | D I V L T Q S P A S L A V S L G Q R A T I S C R A |
| MOPC63 | II | | D I V L T Q S P A S L A V S L G Q R A(T,I,S,C,R) |
| MOPC603 | III | | D I V M T Z S P S S L S V S A G Z ? V T M S C |
| MOPC870 | III | | D I V M T Q S P S S L S V S A G E K V T M S C ? A |
| MOPC384 | III | | D I V M T Q S P S S L S V S A G E K V T M S C |
| MOPC467 | IV | | D V L M T Q T P L S L P V(S)L G D E A ? I(S)C |
| MOPC47 | IV | | E V V M T Q T P L S L A V(S)L G ? Z A(S) |
| MOPC37 | IV | | D V L M |
| MOPC843 | V | | D V V M T Q T P L T L S V T I G E P A ? I S C |
| MOPC674 | V | | D V V M T Q T P L T L S V T I G E P A S L S C |
| MOPC8 | VI | | D I V M T Q S P T F L A V T A S K K V T I S C ? A |
| MOPC15 | VI | | D I V M T Q S P T F L A V T A S K K V T I S C ? A |
| MOPC773 | VII | | E T T V T Q S P A S L S M A I G E K V T I(S,C,R) |
| MOPC265 | VII | | E T T V T Q S P A S L S M A I G E K V T I(S,C,R) |
| LPC 1 | | | D I V M T Q S P S S M Q A S I G E K V T I S C |
| MOPC167 | | | D I V I T Q B E L S D P V T S G E ? V |
| MOPC46 | | | D I V L T Q S P A S L S V T P G D |
| RABBIT | | | |
| 2711 | | | A D V V M T E T P A S V(S)E P V G G P V T I ? C |
| 1305 | | | D V V M T Z T P A T V |
| 2436 | | | A F E L T : T P S S V Z A A V G G ? V T I ? C |
| 2461 | | | A F E M T E T P A S V T A P V |
| | | | I V |
| 2690 | | | - I V M T E T P S S ? S V P V G G T V T I |

(continued)

| PROTEIN | SUB GRP | ALLO-TYPE | 33–70 (3 4 5 6 7 8 9 0 1 2 3 4 5 6 7 8 9 0 1 2 3 4 5 6 7 8 9 0 1 2 3 4 5 6 7 8 9 0) |
|---|---|---|---|
| Hau | I | Val | L S W Y Q Q K P G K A P Q V L I Y A A S S L P S G V P S R F S G S G S G T D |
| Ou | I | Val | L B W Y Z(Z,K,P,G)K A(P,B,L)L I Y A A S B L A S G(V,P,S)R F S G S G S G T B/ |
| Dav | I | | L I W Y Q Q Y P |
| Fin | I | | L I W Y Q Q Y P |
| Eu | I | Val | L A W Y Q Q K P G K A P K L L M Y K A S S L E S G V P S R F I G S G S G T E |
| Paul | I | | L A W Y Z Z K P G Z A P K L L I Y Z |
| HBJ 4 | I | Leu | L A W Y Z E K P G Z A P K L L I S K T S S L E R G V P S R F A G S G |
| Bel | I | Val | G K P P E L L I Y D A S T L K T G V P S R/F S G S G S E T H |
| Au | I | Val | L N W Y Q Q K P G K A P K L L I Y D A S N L E S G V P S R F S G G G S G A H |
| Ag | I | Val | L N W Y Q Q G P K K A P K I L I Y D A S N L E T G V P S R F S G S G F G T D |
| Roy | I | Leu | L N W Y Q Q K P G K A P K L L I Y D A S K L E A G V P S R F S G T G S G T D |
| Ti | II | Val | L A W Y Q Q K P G Q A P R L L I Y V A S S R A T G I P D R F S G S G S G T D |
| Fr 4 | II | Leu | L A W Y Q Q R P G Q A P K A T G I P D R/F S G S G S G(T,D, |
| B6 | II | Val | L A W Y Q Q K P G Q A P R L L M Y G V S S R/A T G I P D R/F S G S G S G A(D, |
| DII | II | Val | L S W Y Z Z K P G Z T P R L L I Y |
| Rad | II | Val | L A W Y Q Q K P G Q A P R A T G T P V R/F S G S E S G T D |
| MII | III | Val | L D W Y/L Z K P G Z S P Z L L I Y/L G S N R/A S G V P N R/F S G S G S G T B |
| Tew | III | Val | L N W Y L Q K P G Q(S,P,Z)L L(P,G=A,L)S N R A S G V P D R F S G S G S G T D |
| Bat | III | | L B ? Y L Z K P G Z(S)P Z L L |
| Cum | III | Val | L N W Y L Q K A G Q Q P S L L I Y T L S Y R A S G V P D R F S G S G S G T D |
| MOUSE | | | |
| MBJ 41 | I | | S B W L Z Z(B,P,G,Z,T)I K R L I Y A T S S L B S G V P K R F S G S R S G S D/ |
| MBJ 70 | II | | M N(W,F,Z)Z K P G Z P P K/L L I Y A A S N Q G S G V P A R/F S G S G S G T D |

| PROTEIN | SUB GRP | ALLO-TYPE | 71–108 (1 2 3 4 5 6 7 8 9 0 1 2 3 4 5 6 7 8 9 0 1 2 3 4 5 6 7 8 9 0 1 2 3 4 5 6 7 8) |
|---|---|---|---|
| Hau | I | Val | F T L T I S S L Q P E D F A T Y Y C Q Q N Y I T P T S F G Q G T R V E I K R |
| Ou | I | Val | F T/F T/I S S/L Z P Z B/F A T Y/Y C Z Z S Y S(S,P)T T F G Z G T R L Z I K R |
| Eu | I | Val | F T L T I S S L Q P D D F A T Y Y C Q Q Y N S D S K M F G Q G T K V E V K G |
| Bel | I | Val | F/T L T I ? S L Z P A B F A T Y Y C Q Q Y B H F P L/T F/G G G T E V E V K |
| Au | I | Val | F T F T I S S L Q P E D I A T Y Y C Q Q Y D Y L P W T F G Q G T K V E I K R |
| Ag | I | Val | F T F T I S G L Q P E D I A T Y Y C Q Q Y D T L P R T F G Q G T K L E I K R |
| Roy | I | Leu | F T F T I S S L Q P E D I A T Y Y C Q Q F D N L P L T F G G G T K V D F K R |
| Ker | I | Leu | Y Y C Q Q Y D D L P P T F G P G T K |
| BJ | I | Leu | Y Y C Q Q Y E N L P Y |
| Ti | II | Val | F T L T I S R L E P E D F A V Y Y C Q Q Y G S S P S T F G Q G T K V E L K R |
| Fr 4 | II | Leu | F)T L T I S R/L E P E D F A V Y F(C,Q,Q,Y)G G S P Q P F G Q G T K/L E I K |
| B6 | II | Val | F,T,L,T,I,S)R/L Z P E D F A V Y/Y C Q Q Y/G S S P F/T F G Q G S K/L E I K |
| Rad | II | Val | F T L T(I,S)R/L E P E D F A V Y Y C Q Q Y E T S P T T F/G Q G T R/L D I K |
| MII | III | Val | F T L K/I S R/V Z A Z B V G V Y Y C/M Q A L Q T P L T F G G G T N V E I K R |
| Tew | III | Val | F T L K I S R V E A E D V G V Y Y C M Z A L Q A P I T F G Q G T R L E I K R |
| Man | III | Val | V E I K |
| Cum | III | Val | F T L K I S R V Q A E D V G V Y Y C M Q R L E I P Y T F G Q G T K L E I R R |
| MOUSE | | | |
| MBJ 41 | I | | Y S L T/I S S L/E S E D F V D Y ? C L Q Y A S S P W T F G G G T K L E I K R |
| MBJ 70 | II | | F S L N I H P M Z Z B B T A M Y F C Z Z S K/E V P W T F G G G T K L E I K/R |

**Table 2-2A.** $V_{\kappa}$-region amino acid sequences. Single-letter code is given in legend to Fig. 3-2.

| PROTEIN | SUB GRP | ALLO-TYPE | 11 12 13 14 14<br>9 0 1 2 3 4 5 6 7 8 9 0 1 2 3 4 5 6 7 8 9 0 1 2 3 4 5 6 7 8 9 0 1 2 3 |
|---|---|---|---|
| Ker | I | Leu | I(F,P,P,S)D(E,Q)L K(S,G,T,A,S)V V C L L |
| BJ | I | Leu | I(F,P,P,S)D(E,Q)L K(S,G,T,A,S)V V C L L |
| HBJ 4 | I | Leu | (T,V,A,A,P,S,V,F,I,F,P,P,S,B,Z,Z,L)K/S,G,T,A,S,V,V)C/L,L,B,B,F,Y,P)R/Z, |
| Roy | I | Leu | T V A A P S V F I F P P S D E Q L K S G T A S V V C L L N N F Y P R E |
| Ag | I | Val | T V A A P S V F I F P P S N E Q L K S G T A S V V C L L N N F Y P R E |
| Eu | I | Val | T V A A P S V F I F P P S D E Q L K S G T A S V V C L L N N F Y P R E |
| Ou | I | Val | T V A A P S V F I F P P S B E Q L K S G T A S V V C L L N N F Y P R E |
| Tew | III | Val | (T,V,A,A,P,S,V,F=I,F,P,P,S,B,Z,Z,L)K(S,G,T,A,S,V)V C L L(N,N,F,Y,P)R(E, |
| MII | III | Val | T(V,A,A,P,S,V,F,I,F,P,P,S,B,Z,Z,L)K/S(G,T,A,S,V,V)C/L L N N F Y P R/E |
| Cum | III | Val | T V A A P S V F I F P P S D E Q L K S G T A S V V C L L N N F Y P R E |
| TI | II | Val | T V A A P S V F I F(P,P,S,D,E)Q L K S G(T,A,S,V,V)C L L B(B,F,Y,P)R E |
| DII | II | Val | (T,V,A,A,P,S,V)F/I,F,P,P,S,B,Z,Z,L)K/S,G,T,A,S,V,V,C/L L N B F Y P R/Z, |
| Au | I | Val | (T,V,A,A,P,S,V,F,I,F,P,P,S,B,Z,Z,L)K/S,G,T,A,S,V,V)C/L,L,N,N,F,Y,P)R/E, |
| Hau | I | Val | T V A A P S V F I F P P S B Z Z L K S G T A S V V C L L B B F Y P R Z |
| Rad | II | Val | I(F,P,P,S)D(E,Q)L K(S,G,T,A S)V V C L L |
| MOUSE | | | |
| MBJ 41 | I | | A B A A P T V S(I,F,P,P,S,S)E Q L T G(G,S,A,S)V V C F L N N F Y P K D |
| MBJ 70 | II | | (A,B,A,A,P,T,V,S,I,F,P,P,S,S,Z,Z,L,T,G,G,S,A,S,V,V)C F(L,B,B,F,Y,P)K/D |

| PROTEIN | SUB GRP | ALLO-TYPE | 14 15 16 17 17<br>4 5 6 7 8 9 0 1 2 3 4 5 6 7 8 9 0 1 2 3 4 5 6 7 8 9 0 1 2 3 4 5 6 7 8 |
|---|---|---|---|
| HBJ 4 | I | Leu | A)K/V,Z,W)K/V,B,B,A,L,Z,S,G,B,S,Z,Z,S,V,T,Z,Z,B,S)K |
| Roy | I | Leu | A K V Q W K V D N A L Q S G N S Q E S V T Z Q D S K D S T Y S L S S T |
| Ag | I | Val | A K V Q W K V D N A L Q S G N S Q E S V T E Q D S K D S T Y S L S S T |
| Eu | I | Val | A K V Q W K V D N A L Q S G N S Q E S V T E Q D S K D S T Y S L S S T |
| Ou | I | Val | A K V Q W K V D N A L Q S G N S Q E S V T E Q D S K D S T Y S L S S T |
| Tew | III | Val | A)K(V,Q,W)K V(D,N,A,L,Q,S,G,N,S,Q,E,S,V,T,E,Q,D,S)K(D,S,T,Y=S,L,S,S,T, |
| MII | III | Val | A K/V Q,W K/V(B,B,A,L,Z,S,G,B,S,Z,Z,S,V,T,Z,Z,B,S)K/D(S,T,Y,S,L,S,S,T, |
| Cum | III | Val | A K V Q W K V D N A L Q S G N S Q E S V T Z Q D S K D S T Y S L S S T |
| TI | II | Val | A K V Q W K V(B,B,A)L(Z,S,G,B,S,Z,Z,S,V,T,Z,Z,B,S)K D(S,T)Y S(L,S,S,T) |
| DII | II | Val | A)K/V,Z,W)K/V,B,B,A,L,Z,S,G,B,S,Z,Z,S,V,T,Z,Z,B,S)K/B,S,T)Y S L(S,S,T) |
| Au | I | Val | A)K/V,Q,W)K/V,D,N,A,L,Q,S,G,N,S,Q,E,S,V,T,E,Q,D,S)K/D,S,T,Y,S,L,S,S,T, |
| Hau | I | Val | A K V Z W K V B B A L Z S G B S Z Z S V T Z Z B S K B S T Y S L S S T |
| MOUSE | | | |
| MBJ 41 | I | | I N V K W K I D G S E R Q B G V L(B,S,B,T,Z,W)D S K D S T Y S M S S T |
| MBJ 70 | II | | I N V K/W K I(D,G,S,E)R(Z,B,G,V,L,B,S,B,T,Z,W,D,S)K/D(S,T,Y,S,M,S,S,T, |
| PIG | | | |
| normal | | | W K V D G V V Q S S G I L D S V T E Q D S K/D S T Y S L S L T |

(continued)

| | SUB | ALLO- | 18 19 20 21 21 |
|---|---|---|---|
| PROTEIN | GRP | TYPE | 9 0 1 2 3 4 5 6 7 8 9 0 1 2 3 4 5 6 7 8 9 0 1 2 3 4 5 6 7 8 9 0 1 2 3 4 |
| Ral | III | Leu | L(Y,A,C,Z,V,T,H,Z,G,L,S,S,P,V,T,K) |
| Gal | III | Leu | L(Y,A,C,Z,V,T,H,Z,G,L,S,S,P,V,T,K) |
| Fr 4 | II | Leu | L Y A C E(V,T,H,Q,G,L,S,S,P,V,T)K S F N R G E C |
| Ker | I | Leu | L Y A C E(V,T,H,Q,G,L,S,S,P,V,T)K S F N R G E C |
| BJ | I | Leu | L Y A C E(V,T,H,Q,G,L,S,S,P,V,T)K S F N R G E C |
| HBJ 4 | I | Leu | (A,B,Y,Z)K/H K/L,Y,A)C/Z,V,T,H,Z,G,L,S,S,P,V,T)K/S,F,B)R/G,E)C |
| Roy | I | Leu | L T L S K A D Y E K H K L Y A C E V T H Q G L S S P V T K S F N R G E C |
| Ag | I | Val | L T L S K A D Y E K H K V Y A C E V T H Q G L S S P V T K S F N R G E C |
| Eu | I | Val | L T L S K A D Y E K H K V Y A C E V T H Q G L S S P V T K S F N R G E C |
| Ou | I | Val | L T L S K A D Y E K H K V Y A C E V T H Q G L S S P V T K S F N R G E C |
| Tew | III | Val | L=T,L,S)K/A,D,Y,E)K/H K/V Y A C(E,V,T,H,Q,G,L=S,S,P,V,T)K S F(N,R,G,E,C) |
| Mil | III | Val | L,T,L,S)K/A(B,Y,Z)K/H K/V(Y,A)C/Z(V,T,H,Z,G,L,S,S,P,V,T)K/S(F,N)R G E C |
| Cum | III | Val | L T L S K A D Y E K H K V Y A C E V T H Q G L S S P V T K S F N R G E C |
| Ti | II | Val | L T L S K A D Y E K H K V Y A C E V(T,H,Q,G)L S(S,P,V)T K S F N R G E C |
| Dil | II | Val | L T L S K/A,B,Y,Z)K/H K/V,Y,A,C,Z,V,T,H,Z,G,L,S,S,P,V,T)K/S,F,B)R/G,Z)C |
| Au | I | Val | L,T,L,S)K/A,D,Y,E)K/H K/V Y A C/E,V,T,H,Q,G,L,S,S,P,V,T)K/S,F,N)R/G,E)C |
| Hau | I | Val | L T L S K A B Y Z K H K V Y A C Z V T H Z G L S S P V T K S F B R G Z C |
| Rad | II | Val | V Y A C E(V,T,H,Q,G,L,S,S,P,V,T)K S F N R G E C |
| Man | III | Val | V Y A C E V T H Q G L S S P V T K S F N R G E C |
| Car(I) | I | | S F N R G E C |
| MOUSE | | | |
| MBJ 41 | I | | L T L T K B Z Y Z R H B S Y T C Z A T H K T S T S P I V K S F N R N E C |
| MBJ 70 | II | | L,T,L,T)K/B(Z,Y,Z)R/H(B,S)Y T C Z(A,T,H)K/T(S,T,S,P,I,V)K/S F N R N E C |
| PIG | | | |
| normal | | | L S L P T S Q Y/L S H N/S Y S C/E V T H K B Z C Z A |
| RABBIT | | | |
| normal | | b4 | S F D R G N C |
| normal | | b5 | F S R K N C |
| normal | | b6 | F S R K S C |
| GUINEA PIG | | | |
| normal | | | T I N R S E C |
| RAT | | | |
| normal | | | N E C |
| CHICKEN, TURKEY, DUCK | | | |
| normal | | | S E C |

**Table 2–2B.** $C_{\kappa}$-region amino acid sequences. Single-letter code is given in legend to Fig. 3–2.

| PROTEIN | SUB GRP | SUB CLASS | 1 2 3 4 5 6 7 8 9 10 1 2 3 4 5 6 7 8 9 20 1 2 3 4 5 6 7 8 9 30 1 2 3 34 |
|---|---|---|---|
| Kern | III | Arg | - Y A L T Q P P S V S V S P G Q T A V I T C S G D N - - - L E K T F |
| X | III | Arg | - Y D L T Q P P S V S V S P G Q T A S I T C S G D K - - - L G D K D |
| Bau | III | Arg | - Y G L T Q P P S L S V S P G Q T A S I T C S G D K - - - L G E Q Y |
| III | III | Lys | - Y E L T Q P P S V S V S P G Q T A S |
| Vin | III | Lys | - Y(V,L=T,Q,P,P,S,V,S,V,A,P,G,Q)M I T C G G D |
| Mil | | | I T C G G D Z |
| Sh | V | Arg | - S E L T Q D P A V S V A L G Q T V R I T C Q G D S - - - L R G Y D |
| Hul | | Arg | Q(S,A,L,S,Z,P,P,S,A,S,G,S,P,G,Z,S,V,T,I,S)?/Z,A,Z,S,- - - I,B/B,S,S, |
| HS 68 | I | Arg | Z S A L T Q P A S V S G S P G Q S I T |
| HS 77 | I | Arg | Z S A L T Q P A S V S G S P G Q S I T |
| HBJ 15 | I | Lys | Z S A L T Q P A S V S G S P G Q T I T |
| HS 70 | I | Arg | Z S A L S Q P A S V S G S P G Q S I T |
| HS 86 | I | Lys | Z S P L A Q P A S V S G S P G E S I T |
| HBJ 8 | I | Lys | Z S A L A Q P A S V S G S P G Q S I T |
| VII | I | Arg | H S A L T Q P A S V S G S L G Q S I T I S C T G T S S D V G G Y N Y |
| Bo | IV | Arg | Z S A L T Q P P(S,A)S G S P G Q S V T I S C T G T S S D V G N D K Y |
| HBJ 2 | IV | Lys | Z S A L T Q P P S A S G S P G Q S V T I S C T G T |
| HS 92 | II | Lys | Z S V L T Q P P S V S G A P G Q R |
| HS 78 | II | Arg | Z S A L T Q P P S V S G T P G Q T V T |
| BJ 98 | II | Arg | Z(S,V,L)T(Z,P,P,S,V)S A A(B,G=Z,A)V T S I C |
| HS 94 | II | Arg | Z S V L T Q P P S V S A A P G Q R |
| New | II | Arg | Z S V L T Q P P S V S A A P G Q K V T I S C S G G S T N - I G N N Y |
| Ha | II | Arg | Z S V L T Q P P S V S G T P G Q R V T I S C S G G S S N G T G N N Y |
| HBJ 7 | II | Arg | Z S V L T Q P P S A S G T P G Q G V T I S C S G S |
| HBJ 11 | II | Arg | Z S V L T Q P P S A S G T P G Q R |
| Koh | II | | Z S V L T Z P P S A S G T P G Z S V T I S |
| Mz | | Arg | A I I S C S G S S S N - M |
| MOUSE | | | |
| MOPC104 | | | Z A V V T Q Q S A L T T S P G E T T V L T C R S S T G A V T T S N Y |
| 5 proteins* | | | Z A(V,V,T,Z,Z,S,A,L/T,T,S,P,G,Z,T,T,V,L/T,C/R/S S(T,G,A,V,T,T,S,B,Y/ |
| RPC 20 | | | Z A V V T(Z,Z,S,A,L/T,T,S,P,G,Z,T,T,V,L/T,C/R/S S(T,G,A,V,T,T,S,B,Y/ |
| H 2020 | | | Z A(V,V,T,Z,Z,S,A,L/T,T,S,P,G,Z,T,T,V,L/T,C/R/S T(T,G,A,V,T,T,G,B,Y/ |
| S 176 | | | Z A(V,V,T,Z,Z,S,A,L/T,T,S,P,G,Z,T,T,V,L/T,C/R/S N(T,G,A,V,T,T,S,B,Y/ |
| S 178 | | | Z A(V,V,T,Z,Z,S,A,L=T,T,S,P,G,Z,T,T,V,L=T,C)R S N T G A V T T S B Y |
| MOPC 315 | | | Z A V T V S N T G A V T T S D Y |
| PIG | | | |
| normal | | | Z T V I - Q E P A M S V S P G G T V T L T C A F S S G S V T T S N Y |
| | | | S V N G T |
| | | | P T S N |
| | | | G |
| CHICKEN | | | |
| normal | | | - - A L T Q P A A V G A Q I ? S T V V |
| | | | V S A L S S V A |
| | | | A L V A P G |
| | | | T |
| DUCK | | | |
| normal | | | - - A L T Q |
| TURKEY | | | |
| normal | | | - A V L T |
| DOG | | | |
| normal | | | Z S V L T (V,T,I,S,C) |
| COW | | | |
| normal | | | Z A S L (V,T,I,S,C) |
| SHEEP | | | |
| normal | | | Z A V L V(T,I,S,C) |

*XP 8, J 698, H 2061, J 558, HOPC I

(continued)

| PROTEIN | SUB GRP | SUB-CLASS | 3 4 5 6 7 (positions 35–72): 5 6 7 8 9 0 1 2 3 4 5 6 7 8 9 0 1 2 3 4 5 6 7 8 9 0 1 2 3 4 5 6 7 8 9 0 1 2 |
|---|---|---|---|
| Kern | III | Arg | V S W F Q Q R P G Q S P L L V I Y H T S E R P S E I P E R F S G S S S G A T |
| X | III | Arg | V C W Y Q Q R P G Q S P V L V I Y Q D N Q R S S G I P E R F S G S N S G N T |
| Bau | III | Arg | V C W Y Q Q K P G Q S P V L V I Y H D S K R P S G I P E R F S G S N S G T T |
| Sh | V | Arg | A A W Y Q Q K P G Q A P L L V I Y G R N N R P S G I P D R F S G S S S G H T |
| Hul | | Arg | V)Y/W,Y,Z,Z,Y,P,G)K A P K L/L I F A(V,S,B,R,S,S,G,I,P,B)R/F S(G,Y)K/S,G,B,S, |
| VII | I | Arg | V S W F Q Q H P G T A P K L I I S E V R N R P S G V S D R F S G S K S A N T |
| Bo | IV | Arg | V S W Y Q Q H P G R A P K L V I F E V S Q R P S G V P D R F S G S K S N D T |
| New | II | Arg | V S W H Q H L P G T A P K L L I Y E D N K R P S G I P D R I S A S K S G T S |
| Ha | II | Arg | V Y W Y Q Q L P G T A P K L L I Y R D D K R P S G V P D R F S G S K S G T S |
| MOUSE | | | |
| MOPC 104 | | | A N W V Q Q P D K H L F T G L I G G T N N R A P G V P A R F S G S L I G N K |
| 5 proteins* | | | A,B,W/V,Z,Z,P,B,K,H,L,F/T,G,L,I,G,G,T,B,B)R/A,P,G,V,P,A)R/F,S,G,S,L,I,G,B)K/ |
| RPC 20 | | | A,B,W/V,Z,Z,P,B,K,H,L,F/T G L L(G,G,T,B,B)R/A,P,G,V,P,A)R/F,S,G,S,L,I,G,B)K/ |
| H 2020 | | | A,B,W/V,Z,Z,P,B,K,H,L,F/T,G,L,I,G,G,T,B,B)R/A,P,G,V,P,A)R/F,S,G,S,L,I,G,B)K/ |
| S 176 | | | A,B,W/V,Z,Z,P,B,K,H,L,F/T,G,L,I,G,G,T,B,B)R/A,P,G,V,P,A)R/F,S,G,S,L,I,G,B)K/ |
| S 178 | | | (A,B,W=V,Z,Z,P,B,K,H,L,F)T G L I G N T B B R(A,P,G,V,P,A)R(F,S,G,S,L,I,G,B)K |
| MOPC 315 | | | (A,S)W I E E P D K H L F T G L I G G T S N R |
| PIG | | | |
| normal | | | P G W F Q Q T P G Q P P R |

| PROTEIN | SUB GRP | SUB-CLASS | 7 8 9 10 11 (positions 73–112): 3 4 5 6 7 8 9 0 1 2 3 4 5 6 7 8 9 0 1 2 3 4 5 6 7 8 9 0 1 2 3 4 5 6 7 8 9 0 1 2 |
|---|---|---|---|
| Mz | | | Y C A T W R |
| HS 5 | | Lys | (S, |
| Kern | III | Arg | A T L T I S G A Q S V D E A D Y F C Q T W D T I T - - A I F G G G T K L T V L S |
| X | III | Arg | A T L T I S G T Q A M D E A D Y Y C Q A W D S M S - - V V F G G G T R L T V L S |
| Bau | III | Arg | A T L T I S G T Q A M D E A D Y Y C Q A W D S Y T - - V I F G G G T K L T V L G |
| Sh | V | Arg | A S L T I T G A Q A E D E A D Y Y C N S R D S S G K H V L F G G G T K L T V L G |
| Hul | | Arg | A,S,L)T(I,S,G,L/Q Q E D E A T Y Y/C,S,S,K)Y/T,L=S,T,S)V V F(G,G,G,T)K L(T,V,L,G, |
| VII | I | Arg | A S L T I S G L Q A E D E A D Y Y C S S Y T S S N S - V V F G G G T K L T V L G |
| Bo | IV | Arg | A S L T V S G L R A E D E A D Y Y C S S Y V D N N N F(V)V F G G G T K L T V L R |
| HS 92 | II | Lys | (G, |
| BJ 98 | II | Arg | L(T,V,L,G, |
| New | II | Arg | A T L G I T G L R T G D E A D Y Y C A T W D S S L N A V V F G G G T K V T V L G |
| Ha | II | Arg | A S L A I S G L R S E D E A H Y H C A A W D Y R L S A V V F G G G T Q L T V L R |
| MOUSE | | | |
| MOPC 104 | | | A A L T I T G A Q T E D E A I Y F C A L W Y S N H - - W V F G G G T K L T V L G |
| 5 proteins* | | | A,A,L/T,I,T,G,A,Z,T,Z,B,Z,A,I,Y)F C/A,L,W/Y,S,B,H,- - W/V,F,G,G,G,T)K/L,T,V,L/G, |
| RPC 20 | | | A,A,L/T,I,T,G,A,Z,T,Z,B,Z,A,I,Y)F C/A,L,W/Y,S,B,H,- - W/V,F,G,G,G,T)K/L,1,V,L/G, |
| H 2020 | | | A,A,L/T,I,T,G,A,Z,T,Z,B,Z,A,I,Y)F C/A,L,W/Y,S,B,H,- - W/V,F,G,G,G,T)K/L,T,V,L/G, |
| S 176 | | | A,A,L/T,I,T,G,A,Z,T,Z,B,Z,A,I,Y)F C/A,L,W/Y,S,B,H,- - W/V,F,G,G,G,T)K/L,T,V,L/G, |
| S 178 | | | (A,A,L=T,I,T,G,A,Z,T,Z,B,Z,A,I,Y)F C(A,L,W=Y,S,B)R - - W(V,F,G,G,G,T)K(L,T,V,L=G, |
| PIG | | | |
| normal | | | A T L T I T G A Q A E D E A D Y F C A L Y K |
| | | | A L A N S G R |
| | | | G G T |
| | | | G |
| BABOON, MONKEY, GUINEA PIG | | | |
| normal | | | (L,T,V,L,G, |
| SHEEP | | | |
| normal | | | L T(V,L,G, |

*XP 8, J 698, H 2061, J 558, HOPC I

**Table 2-3A.** $V_{\lambda}$-region amino acid sequences. Single-letter code is given in legend to Fig. 3-2.

| PROTEIN | SUB GRP | SUB CLASS | 113–166 (3 4 5 6 7 8 9 0 1 2 3 4 5 6 7 8 9 0 1 2 3 4 5 6 7 8 9 0 1 2 3 4 5 6 7 8 9 0 1 2 3 4 5 6 7 8 9 0 1 2 3 4 5 6) |
|---|---|---|---|
| HS 92 | II | Lys | Q,P)K/A(A,P S,V,T,L,F,P,P,S,S,Z,Z,L,Z,A,B)K/A(T,L,V)C/L(I,S,D,F,Y,P,G,A,V,T,V,A,W)K/A(D,S,S,P,V)K/A(G,V,E,T, |
| HS 5 | | Lys | Q,P)K/A(A,P,S,V,T,L,F,P,P,S,S,Z,Z,L,Z,A,B)K/A(T,L,V)C/L(I,S,D,F,Y,P,G,A,V,T,V,A,W)K/A(D,S,S,P,V)K/A(G,V,E,T, |
| Kern | III | Arg | Q P K A A P S V T L F P P S S E E L Q A N K A T L V C L I S D F Y P G A V T V A W K A D G S P V K A G V E T |
| Vil | I | Arg | Q P K A A P S V T L F P P S S E E L Q A N K A T L V C L I S D F Y P G A V T V A W K A D S S P V K A G V E T |
| Sh | V | Arg | Q P K A A P S V T L F P P S S E E L Q A N K A T L V C L I S D F Y P G A V T V A W K A D S S P V K A G V E T |
| Ha | II | Arg | Q P K A A P S V T L F P P S S E E L Q A N K A T L V C L I S D F Y P G A V T V A W K A D S S P V K A G V E T |
| Bo | IV | Arg | Q P K A A P S V T L F P P S S E E L Q A N K A T L V C L I S D F Y P G A V T V A W K A D S S P V K A G V E T |
| New | II | Arg | Q P K A A P S V T L F P P S S E E L Q A N K A T L V C L I S D F Y P G A V T V A W K A D S S P V K A G V E T |
| X | III | Arg | Q P K A A P S V T L F P P S S E E L Q A N K A T L V C L I S D F Y P G A V T V A W K A D S S P V K A G V E T |
| Bau | III | Arg | Q P K A(A,P S,V,T,L,F,P,P,S,S,Z,Z,L,Z,A,B)K A T L V C L(I,S,D,F,Y,P,G,A,V,T,V,A)W K A D S S P V K A(G,V,E,T, |
| Mz | | Arg | (Q,P)K(A,A,P,S,V,T,L,F,P,P,S,S,Z,Z,L,Z,A,B)K A(T,L,V,C,L,I,S,D,F,Y,P,G,V V T)V A W K(A,D,S,S,P,V)K(A,G,V,E,T, |
| Hul | | Arg | Z P)K/A,A,P,S,V,T/L,F,P,P,S,S,Z,Z,L,Z,A)B K(A,T,L/V,C)L (V,A)W K A(D,S,S,P,V)K/A,G,V,Z,T, |
| BJ 98 | II | Arg | Q,P)K/A,A,P,S,V,T,L,F,P,P,S,S,E,E,L,Q,A,N)K/A T L V C |
| 9 proteins* | | | (A,T,L,V,C,L,I,S,D,F,Y,P,G,A,V,T,V,A,W,K) |
| MOUSE | | | |
| MOPC 104 | | | Q P K S S P S V T L F P P S S E E L T E N K A T L V C T I T D F Y P G V V T V D W K V D G T P V T Q G M E T |
| 9 proteins** | | | Z,P)K |
| MONKEY normal | | | Z,P)K (A,T,L,V)C |
| BABOON normal | | | Z,P)K (A,T,L,V)C |
| GUINEA PIG normal | | | Z,P)K (A,T,V,V)C |
| DOG normal | | | A(T,L,V)C |
| PIG normal | | | (A,T,L)V C L |
| SHEEP normal | | | Z,P)K A(T,V,V)C |
| COW normal | | | (A,T,L,V)C |
| CHICKEN normal | | | (A,T,L,V,C) |
| TURKEY normal | | | (A,T,L,V,C) |

*Mil, Fr, Sch, Wil, Bu, We, Li, Lw, Lb
**XP 8, J 698, H 2061, J 558, HOPC 1, S 176, H 2020, S 178, RPC 20

(continued)

```
           SUB SUB-          17                  18                  19                  20                  21            21
PROTEIN    GRP CLASS    7 8 9 0 1 2 3 4 5 6 7 8 9 0 1 2 3 4 5 6 7 8 9 0 1 2 3 4 5 6 7 8 9 0 1 2 3 4 5 6 7 8 9 0 1 2 3 4 5 6 7
Vin        III Lys                                                        K                                     P T E C S
HS 92      II  Lys      T,T,P,S)K/Q(S,N,N)K/Y(A,A,S,S,Y,L,S,L,T,P,Z,Z,W)K/S H K/S(Y,S)C/Z(V,T,H,Z,G,S,T,V,Z)K/T(V,A,P,T,E,C,S)
HS 5           Lys      T,T,P,S)K/Q(S,N,N)K/Y(A,A,S,S,Y,L,S,L,T,P,Z,Z,W)K/S H K/S(Y,S)C/Z(V,T,H,Z,G,S,T,V,Z)K/T(V,A,P,T,E,C,S)
Kern       III Arg      T T P S K Q S N N K Y A A S S Y L S L T P E Q W K S H R S Y S C Q V T H E G S T V E K T V A P T E C S
Vil        I   Arg      T T P S K Q S N N K Y A A S S Y L S L T P E Q W K S H R S Y S C Q V T H E G S T V E K T V A P T E C S
Sh         V   Arg      T T P S K Q S N N K Y A A S S Y L S L T P E Q W K S H R S Y S C Q V T H E G S T V E K T V A P T E C S
Ha         II  Arg      T T P S K Q S N N K Y A A S S Y L S L T P E Q W K S H R S Y S C Q V T H E G S T V E K T V A P T E C S
Bo         IV  Arg      T T P S K Q S N N K Y A A S S Y L S L T P E Q W K S H R S Y S C Q V T H E G S T V E K T V A P T E C S
New        II  Arg      T T P S K Q S N N K Y A A S S Y L S L T P E Q W K S H R S Y S C Q V T H E G S T V E K T V A P T E C S
X          III Arg      T T P S K Q S N N K Y A A S S Y L S L T P E Q W K S H R S Y S C Q V T H E G S T V E K T V A P T E C S
Bau        III Arg      T,T,P,S)K(Z,S,B B)K Y(A,A,S,S,Y,L,S,L,T,P,Z,Z)W K S H R S Y S C(Z,V,T,H,Z,G,S,T,V,Z)K T(V,A,P,T,E)C S
Mz             Arg      T,T,P,S)K(Q,S,N,N,N,Y,A,A,S,S,Y,L,S,L,T,P,E,Q,W)K(S,H)R(S,Y,S,C,Z,V,T,H,Z,G,S,T,V,Z)K(T,V,A,P,T,E,C,S)
Hul            Arg      T,T,P,S)K/Q(S,B,B)K Y(A,A,S,S)Y L S L(T,P,Z,Z,W)K/S H R/S,Y,S,C,Z/V,T,H,Z,G,S,T/V,Z)K/T V A(P,T,Z,C)S
Daw                                                                                                       (K,T,V,A,P,T,E,C,S)
3 proteins*                                (Y,A,A,S,S,Y,L,S,L,T,P,E,Q,W,K)
6 proteins**                     (Q,S,N,N,K=Y,A,A,S,S,Y,L,S,L,T,P,E,Q,W,K)
MOUSE
MOPC 104                T E P S K Q S N N K Y M A S S Y L T L T R A W E R S H S Y S S C Q V T(H,Z,G,H,T)V Q K S L S R A D C S
RPC 20                                                                                                      K S L S R A D C S
MOPC 315                                                                                                                  C L
CHIMP normal                                                              S H R
                                                                          S H K
MONKEY normal                    (Q,S,N,N)K                               S H K        (Z,V,T,H,Z,G,S,T,V,Z)K
BABOON normal                    (Q,S,N,N)K                               S H K        (Z,V,T,H,Z,G,S,T,V,Z)K
RABBIT normal                                                                                             G B(G,B,A,P,T,E,C)S
GUINEA PIG normal                (Q,S,N,N)K                                                                   S L A P S E C S
DOG normal                       (Q,S,N,N)K                                   (S,F,H)C/L V T(H,Z,G,S,T,V,Z)K    V A P A E C S
HORSE normal                                                                                                    L S P S E C P
PIG normal                                                                        F T C Q V T H E G T         T V T P S E C A
SHEEP normal                     (Q,S,N,N)K                                     S(Y,T)C/Z(V,T,H,Z,G,S,T,V,Z)K
COW normal                       (Q,S,N,N)K                                  (G,S,Y,S)C/Z,V,T,H,Z,G,S,T,V,Z)K/T V K P S G C S
```

*Mil, Fr, Sch
**Wil, Bu, We, Li, Lw, Lb

**Table 2–3B.** $C_\lambda$-region amino acid sequences. Single-letter code is given in legend to Fig. 3–2.

```
         SUB (SUB)                    1                   2                   3
PROTEIN  GRP CLASS 1 2 3 4 5 6 7 8 9 0 1 2 3 4 5 6 7 8 9 0 1 2 3 4 5 6 7 8 9 0 1 2 3 a
Ste      I   γ1    Z V H L V E S S A E V K K P G A S M K V S C R A
Di       I   μ     Z V Q L(T,Z,S,G,A,G,L)K K(P,G,Z,P -)K
Ca       I   γ1    Z V Q L V Q S G A E V R K P G A S V K I S C K T S G Y T F S H Y A M
Eu       I   γ1    Z V Q L V Q S G A E V K K P G S S V K V S C K A S G G T F S R S A -
Dee      I   γ1                                      V R I(S,C,K,A,S,G)
He       II  γ1    Z V T L K E N G P T L V K P T E T L T L T C T L S G L S L T T D G V
Ou       II  μ     Z V T L T E S G P A L V K P K Q P L T L T C T F S G F S L S T S R M
Daw      II  γ1    Z V T L R E S G P A L V R P T Q T L T L T C T F S G F S L S G E T M
Cor      II  γ1    Z V T L R E S G P A L V K P T Q T L T L T C T F S G F S L S S T G M
Nie      III γ1    Z V Q L V Q S G G G V V Q P G R S L R L S C A A S G F T F S R Y T -
Vin      III γ4    E V Q L V E S G G G L I Q P G G S L R L S C A A S G F T V S T N Y M
Hin      III γ     G V L L V E S G G V S I ? P G G S L ? L ? C ? A S G F ? I G ? F ? M
Eik      III γ1    E V Q L V E S G G G L V ? P G G S L R L S C A T T G F A F S G S A V
Til      III γ2    E V Q L L E S G G G L V Q P G G S L R L S C A A S G F
Til      III μ     E V Q L L E S G G G L V Q P G G S L R L S C A A S G F
For      III α1    E I Q L V E S G G G L V K G G G S L R L S C A A S G F
Wat      III γ2    E V Q L V E S G G G L V Q P G G S L R L S C A A S G F
Wo       III μ     E V Q L V E S G G G L V Q P G G S L R L
Zuc      III γ3    Z V Q V V E S G A D L V K P G G S S - - - - - - - - - - - - - -(to 216)
Ha       III α1    E V Q L V E S G G G L V G P G
Na       III μ     E V Q L V E S G G A L
MOUSE
MOPC173      γ2a   E V K L L E S G G P L V Q L G G S L K L S C A A S G F D F S R Y W M
RABBIT
normal γ (a1)      Z - S L E E S G G R L V T P T P G L T L T C T A S G F S L S S Y A M
                         V                                       V                 D
normal γ (a3)      Z - S L E E S G G D L V K P G A S L T L T C T A S G F N A S S Y Y M
                                                                           G     F
normal γ (a3)      Z E Q L(E,E,S,G,G,V,L=V,?,P,G,?,S,L)T L T C
                                     D
```

(continued)

| | SUB | (SUB) | | 3 | | | | | | 4 | | | | | | | | | | | 5 | | | | | | | | | | | 6 | | | | |
|---|---|---|---|---|---|---|---|---|---|---|---|---|---|---|---|---|---|---|---|---|---|---|---|---|---|---|---|---|---|---|---|---|---|---|---|---|
| PROTEIN | GRP | CLASS | b | 4 | 5 | 6 | 7 | 8 | 9 | 0 | 1 | 2 | 3 | 4 | 5 | 6 | 7 | 8 | 9 | a | 0 | 1 | 2 | a | 3 | 4 | 5 | 6 | 7 | 8 | 9 | 0 | a | 1 | 2 | 3 |
| Eu | I | γ1 | - | I | I | W | V | R | Q | A | P | G | Q | G | L | E | W | M | G | - | G | I | V | - | P | M | F | G | P | P | N | Y | - | A | Q | K |
| He | I | γ1 | A | V | G | W | I | R | Q | G | P | G | R | A | L | E | W | L | A | W | L | L | Y | W | D | - | - | D | D | K | R | F | - | S | P | S |
| Ou | II | μ | R | V | S | W | I | R | R | P | P | G | K | A | L | E | W | L | A | - | R | I | B | - | B | - | - | B | D | K | F | Y | W | S | T | S |
| Daw | II | γ1 | C | V | A | W | I | R | Q | P | P | G | E | A | L | E | W | L | A | W | D | I | L | - | N | - | - | D | D | K | Y | Y | - | G | A | S |
| Cor | II | γ1 | C | V | G | W | I | R | Q | P | P | G | K | G | L | E | W | L | A | - | R | I | D | W | D | - | - | D | D | K | Y | Y | - | B | T | S |
| Nie | III | γ1 | - | I | H | W | V | R | Q | A | P | G | K | G | L | E | W | V | A | - | V | M | S | - | Y | B | G | B | B | K | H | Y | - | A | D | S |

| | | | 6 | | | | | | 7 | | | | | | | | | | 8 | | | | | | | | | | 9 | | | | | | | |
|---|---|---|---|---|---|---|---|---|---|---|---|---|---|---|---|---|---|---|---|---|---|---|---|---|---|---|---|---|---|---|---|---|---|---|---|---|
| | | | 4 | 5 | 6 | 7 | 8 | 9 | 0 | 1 | 2 | 3 | 4 | 5 | 6 | 7 | 8 | 9 | 0 | 1 | 2 | 3 | 4 | 5 | 6 | 7 | 8 | 9 | 0 | 1 | 2 | 3 | 4 | 5 | 6 | 7 |
| Eu | I | γ1 | F | Q | G | R | V | T | I | T | A | D | E | S | T | N | T | A | Y | M | E | L | S | S | L | R | S | E | D | T | A | F | Y | F | C | A |
| Dee | I | γ1 | | | | | | | | | | | | | | | | | | | | | | | | | | | | | | | Y | Y | C | T |
| He | II | γ1 | L | K | S | R | L | T | V | T | R | D | T | S | K | N | Q | V | V | L | T | M | T | N | M | D | P | V | D | T | A | T | Y | Y | C | V |
| Ou | II | μ | L | R | T | R | L | S | I | S | K | N | D | S | K | N | Q | V | V | L | I | M | I | N | V | N | P | V | D | T | A | T | Y | Y | C | A |
| Daw | II | γ1 | L | E | T | R | L | A | V | S | K | D | T | S | K | N | Q | V | V | L | S | M | N | T | V | G | P | G | D | T | A | T | Y | Y | C | A |
| Cor | II | γ1 | L | E | T | R | L | T | I | S | K | D | T | S | R | N | Q | V | V | L | T | M | - | - | - | D | P | V | D | T | A | T | Y | Y | C | A |
| Nie | III | γ1 | V | N | G | R | F | T | I | S | R | N | D | S | K | N | T | L | Y | L | N | M | N | S | L | R | P | Z | B | T | A | V | Y | Y | C | A |
| RABBIT normal | | γ | | | | | | | | | | | | | | | | | | | | | | | I | T | S | P | T | Q | D | T | A | T | Y | F | C | A |

| | | | | | | 10 | | | | | | | | | | | 10 | | | | | | | 11 | | | | | | | 11 |
|---|---|---|---|---|---|---|---|---|---|---|---|---|---|---|---|---|---|---|---|---|---|---|---|---|---|---|---|---|---|---|---|---|
| | | | a | 8 | 9 | 0 | 1 | 2 | a | b | c | d | e | f | g | h | i | 3 | 4 | 5 | 6 | 7 | 8 | 9 | 0 | 1 | 2 | 3 | 4 | 5 | 6 | 7 |
| Eu | I | γ1 | - | G | G | Y | G | I | - | - | - | - | - | - | - | - | - | Y | S | P | E | E | Y | N | G | G | L | V | T | V | S | S |
| Dee | I | γ1 | - | G | R | G | M | | | | | | | | | | | | | | | | | | | | | | | | | |
| He | II | γ1 | H | R | H | P | R | T | L | - | - | - | - | - | - | - | - | A | F | D | V | W | G | Q | G | T | K | V | A | V | S | S |
| Ou | II | μ | - | R | V | V | N | S | V | M | - | A | G | Y | Y | Y | Y | Y | M | D | V | W | G | K | G | T | T | V | T | V | S | S |
| Daw | II | γ1 | - | R | S | C | G | S | Q | - | - | - | - | - | - | - | - | Y | F | D | Y | W | G | Q | G | I | L | V | T | V | S | S |
| Cor | II | γ1 | - | R | I | T | V | I | P | A | P | A | G | - | - | - | - | Y | M | D | V | W | G | R | G | T | P | V | T | V | S | S |
| Nie | III | γ1 | - | R | I | R | D | T | A | M | - | - | - | - | - | - | - | F | F | A | H | W | G | Q | G | T | L | V | T | V | S | S |
| Vin | III | γ4 | | | | | | | | | | | | | | | | | | | | | | | | | | V | T | V | S | S |
| RABBIT normal | | γ | - | R | | | | | | | | | | | | | | | | | | | B | L(G,G,L,V,T) | | | | | | V | S | Z P |

**Table 2–4A.** $V_H$-region amino acid sequences. Single-letter code is given in legend to Fig. 3–2.

| PROTEIN | SUB GRP | (SUB) CLASS | ALLO-TYPE | 11 12 13 14 15<br>8 9 0 1 2 3 4 5 6 7 8 9 0 1 2 3 4 5 6 7 8 9 0 1 2 3 4 5 6 7 8 9 0 1 2 |
|---|---|---|---|---|
| Eu | I | γ1 | 4,22 | A S T K G P S V F P L A P S S K S T S G G T A A L G C L V K D Y F P E |
| He | III | γ1 | 4,22 | A S T K G P S V F P L A(P,S,S,K) |
| Cor | II | γ1 | 1,17 | A S T K G P S V F P L A P S S K S T S G G T A A L G C L V K D Y F P E |
| Daw | II | γ1 | 1,17 | A S T K G P S V F P L A P S S K S T S G G T A A L G C L V K D Y F P E |
| Car | | γ1 | 1,17 | (S,T,S,G,G,T,A,A)L G C L V K |
| Dee | I | γ1 | 1,17 | (S,T,S,G,G,T,A,A)L G C L V K |
| Nie | III | γ1 | 1,17 | A S T K |
| Sa | | γ2 | | P L A P C S R S T S E S T/A A L G C L |
| Kup | | γ3 | | P L A P C S R(S,T,S,G,G,A,A,A,L/G C L |
| Bru | | γ3 | | P L A P C S R(S,T,S,G,G,A,A,A,L/G C L |
| Vin | III | γ4 | | A S T K G P S V F P L A P C S R S T S E S T P A L G C L V K |
| RABBIT normal | | γ | | S G T K A P S V F P L A P C C G D T P S S T V T L G C L V K G Y L P E |
| G.PIG normal | | γ2 | | T(T,A,P,S,V,F,P,L)A(V,C,S,G,B,T,A)M S T L S M G(C,L,V)K G Y F P Z |
| MOUSE | | | | |
| MOPC 173 | | γ2a | | P V C G D T T G S S |
| MP 5563 | | γ2a | | P V C G D T S G S S |
| MOPC 141 | | γ2b | | T T P P S V Y P L A P G C G |
| HUMAN | | | | |
| Ou | II | μ | | G S A S A P T L F P L V S C E N S(B,P,S,S,T)V A V G C L A Z D F L P D |
| Ale | | μ | | P L V S C Z B S(D,P,S,S,T) |
| Er | | δ | | P I I S G C R |

| PROTEIN | SUB GRP | (SUB) CLASS | ALLO-TYPE | 15 16 17 18 18<br>3 4 5 6 7 8 9 0 a 1 2 3 4 5 6 7 8 9 0 1 2 3 4 5 6 7 8 9 0 1 2 3 4 5 6 7 |
|---|---|---|---|---|
| Eu | I | γ1 | 4,22 | P V T V S W N S - G A L T S G V H T F P A V L Q S S G L Y S L S S V V T |
| Cor | II | γ1 | 1,17 | P V T V S W N S - G A L T S G V H T F P A V L Q S S G L Y S L S S V V T |
| Daw | II | γ1 | 1,17 | P V T V S W N S - G A L T S G V H T F P A V L Q S S G L Y S L S S V V T |
| RABBIT normal | | γ | | P V T V T W N S - G T L T D G V R T F P S V R Q S S G L Y S V P S T V S |
| G.PIG normal | | γ2 | | (P,V,T,V)K |
| HUMAN | | | | |
| Ou | II | μ | | S I T F S W K Y(B,B,S,B,K)I SST R G F P S V L R - G G K Y A A(Z,S,T)V L |

| PROTEIN | SUB GRP | (SUB) CLASS | ALLO-TYPE | 19 20 21 21<br>8 9 0 1 a b 2 3 4 5 6 7 8 9 0 1 2 3 4 5 6 7 8 9 0 1 2 3 4 a b c d e f 5 6 |
|---|---|---|---|---|
| Eu | I | γ1 | 4,22 | V P S S - - S L G T Q T Y I C N V N H K P S N T K V D K R - - - - - - - V E |
| Daw | II | γ1 | 1,17 | V P S S - -(S)L G T Q T Y I C N V N H(K,P,S)N T K V D K K - - - - - - - V E |
| Cor | II | γ1 | 1,17 | V P S S - -(S)L G T Q T Y I C N V N H(K,P,S)N T K V D K K - - - - - - - V E |
| Car | | γ1 | 1,17 | Y I C N V N H K(P,S,N,T/K(V,D)K K - - - - - - - V(E, |
| Dee | I | γ1 | 1,17 | Y I C N V N H K(P,S,N,T/K(V,D)K K - - - - - - - V(E, |
| Sa | | γ2 | | Y(T,C,B,V,B,H,K,P,S,B,T/K V B K T - - - - - - - V Z |
| Zuc | III | γ3 | 11 | (from 18) - - - - - - - - - - - - - - - - - - - - - - - - - - - - E |
| Kup | | γ3 | | (Y,T,C,B,V,B,H,K,P,S)K T P L G D T(C,T,P,H)R C P E |
| Bru | | γ3 | | (Y,T,C,B,V,B,H,K,P,S,K) |
| Vin | III | γ4 | | T Y T C N V D H K P S N T K V D K R - - - - - - - V E |
| RABBIT normal | | γ | | V S Z P - -(P,S)- - - - - T C B V A H - A T B T K V D K T - - - - - - - V A |
| G.PIG normal | | γ2 | | V P S S - - E K - - - - - - T C N V A H P A S T S K V D K T - - - - - - - V E |
| MOUSE | | | | |
| MOPC 21 | | γ1 | | D C G |
| MOPC 173 | | γ2a | | G P T I K P - - - - - - C C P |
| MP 5563 | | γ2a | | G P T I K P - - - - - - C C P |
| MOPC 141 | | γ2b | | I N P - - - - - - C P P |
| HUMAN | | | | |
| Ou | | μ | | L P S K D V M Q G(B,Z,T,H)V C K V D H R - - - - - - - G L |

(continued)

| PROTEIN | SUB GRP | (SUB) CLASS | ALLO-TYPE | 21 22 23 24 25<br>7 8 9 0 1 2 3 4 5 6 7 8 9 0 1 2 3 4 5 6 7 8 9 0 1 2 3 4 5 6 7 8 9 0 1 2 |
|---|---|---|---|---|
| Eu | I | γ1 | 4,22 | P K S C D K T H T C P P C P A P E L L G G P S V F L F P P K P K D T L M |
| Cor | II | γ1 | 1,17 | P K S C D K T H |
| Daw | II | γ1 | 1,17 | P K S C D K T H |
| Car | | γ1 | 1,17 | P)K S C D K T H T C P P C P A P E L(L,G,G,P,S,V,F,L,F,P,P,K,P)K |
| Dee | | γ1 | 1,17 | P)K S C D K T H T C P P C P A P E L(L,G,G,P,S,V,F,L,F,P,P,K,P)K |
| Sa | | γ2 | | R K C C V Z - - - C P P C P A(P,G,A,V)S F |
| Zuc | III | γ3 | 11 | P K S C D T P P P C P R C P A P E L L G G P S V F L F P P K P K D T L M |
| Kup | | γ3 | | P K/S C D T P P P C P R C P A P E L |
| Bru | | γ3 | | P K/S C D T P P P C P R C P A P E L |
| Vin | III | γ4 | | S K - - - Y G P P C P P C P A(S,E)F L G G P S V F L F P P K P K D T L M |
| RABBIT normal | | γ | | P S T C S K - P T C P P - - - P E L L G G P S V F I F K P P P K D T L M<br>M |
| G.PIG normal | | γ2 | | P I R T P Z P(B,S/C P P - - - P E B/L/G G P S V F I F P P K P K D T L M |
| MOUSE | | | | |
| MOPC 21 | | γ1 | | C K P C I - - - - - - - - C T V P E V - - - S S V F I F P P K P K |
| MOPC 173 | | γ2a | | P K - - - - - - - - - - - C P A P N L |
| MP 5563 | | γ2a | | P K - - - - - - - - - - - C P A P N L |
| MOPC 141 | | γ2b | | C K E C H K - - - - - - - C P A P N L |
| HUMAN | | | | |
| Ou | II | μ | | T F Z Z B A S S M C - - - - V P D Q D T A I R V F A I P P S F A S I F L |
| Er | | δ | | T P E C P S H T Q P L G V |

| PROTEIN | SUB GRP | (SUB) CLASS | ALLO-TYPE | 25 26 27 28 28<br>3 4 5 6 7 8 9 0 1 2 3 4 5 6 7 8 9 0 1 2 3 4 5 6 7 8 9 0 1 2 3 4 5 6 7 8 |
|---|---|---|---|---|
| Cra | | γ1 | 4,22 | I S R T P E(V,T)C V V(V,D) |
| Eu | I | γ1 | 4,22 | I S R T P E V T C V V V D V S H E D P Q V K F N W Y V D G V Q V H N A K |
| Sa | | γ2 | | I S R(T,P,E,V,T,C,V) |
| Zuc | III | γ3 | 11 | I S R T P E V T C V(V,V,D) |
| Vin | III | γ4 | | I S R T P E V T C V V V D V S Q E D P Z(V,Z,F)N W Y V D G V E V H N A K |
| NORMAL rabbit | | γ | | I S R T P E V T C V V V D V S E D D P E(V,Z)F T W Y I B B Z Z V R/T A R |
| HUMAN | | | | |
| Ou | II | μ | | T K S T K L |

| PROTEIN | SUB GRP | (SUB) CLASS | ALLO-TYPE | 29 30 31 32<br>9 0 1 2 3 4 5 6 7 8 9 0 1 2 3 4 5 6 7 8 9 0 1 2 3 4 5 6 7 8 9 0 1 2 3 4 |
|---|---|---|---|---|
| Cra | | γ1 | 4,22 | Y K C K V S |
| Eu | I | γ1 | 4,22 | T K P R E Q Q Y B S T Y R V V S V L T V L H Q N W L D G K E Y K C K V S |
| Sa | | γ2 | | T K C K V S |
| Zuc | III | γ3 | 11 | Y K C K V S |
| Vin | III | γ4 | | T K P R E E Q F B S T Y R/V V(S,V,L=T,V,L,H,Z,B,W,L,D,G)K/E Y K C K V S |
| RABBIT normal | | γ | | P P L R E Q Q F D S T I R V V S T L P I T H E D W L R G K E F K C K V H<br>A |

| PROTEIN | SUB GRP | (SUB) CLASS | ALLO-TYPE | 32 33 34 35 36<br>5 6 7 8 9 0 1 2 3 4 5 6 7 8 9 0 1 2 3 4 5 6 7 8 9 0 1 2 3 4 5 6 7 8 9 0 |
|---|---|---|---|---|
| Cra | | γ1 | 4,22 | N K A L P A P I E ... E E M T K |
| Eu | I | γ1 | 4,22 | N K A L P A P I E K T I S K A K G Q P R E P Q V Y T L P P S R E E M T K |
| unnamed(1) | | γ1 | 1,17 | T L P P S R D E L T K |
| unnamed(2) | | γ1 | 1,17 | (D,E,L,T)K |
| Sa | | γ2 | | N K G L P A P I E |
| Zuc | III | γ3 | 11 | N K A L P A P I E |
| Vin | III | γ4 | | N K G L P S S I E K T I S K A K G Q P R E P Q V Y T L P P S Q E E M T K |
| RABBIT normal | | γ | | D K A L P A P I E K T I S K A R G E P L E P K V Y T M G P P R E Q L S S |
| MONKEY normal | | γ | | E E L T K |

(continued)

**Table 2-4B.** $C_H$-region amino acid sequences. Single-letter code is given in legend to Fig. 3-2.

Table 2–4B (continued)

| PROTEIN | SUB GRP | (SUB) CLASS | ALLO-TYPE | 36 1 2 3 4 5 6 7 8 9 37 0 1 2 3 4 5 6 7 8 9 38 0 1 2 3 4 5 6 7 8 9 39 0 1 2 3 4 5 6 7 8 9 40 0 1 2 3 4 5 6 7 8 9 41 0 1 2 3 4 5 |
|---|---|---|---|---|
| Cra | | γ1 | 4,22 | (L,T,C,L) |
| Eu | I | γ1 | 4,22 | N Q V S L T C L V K G F Y P S D I A V E W E S N D G E P E N Y K T T P P V L D S D G S F F L Y S K L T V D K S |
| Sa | | γ2 | | T(C,L) |
| Zuc | III | γ3 | 11 | L T(C,L) |
| Vin | III | γ4 | | N Q V S L T C L V K G F Y P S D I A V E W Z S(B,B,G,Z,P,Z,B,Y)K/T T P P(V,L)D S D G S F(F)L Y S R L T V D K S |
| RABBIT | normal | γ | | R S V S L T C M I D G F Y P S D I S V G W E K D G K A E D D Y K T T P A V L D S D G S W F L Y S K L S V P T S |
| MOUSE | | | | |
| MOPC 173 | | γ2a | | C M V T B F M P Z B I Y V Z |
| HUMAN | | | | |
| Ou | II | μ | | M Q R G E P L S P Q K Y V T S A P M P E P Q A P G R Y F A H S I L T V S E E |

| PROTEIN | SUB GRP | (SUB) CLASS | ALLO-TYPE | 41 6 7 8 9 42 0 1 2 3 4 5 6 7 8 9 43 0 1 2 3 4 5 6 7 8 9 44 0 1 2 3 4 5 6 7 8 9 45 0 1 2 3 4 5 6 7 8 9 46 0 1 2 3 4 5 |
|---|---|---|---|---|
| Cra | | γ1 | 4,22 | F S C S V M |
| Eu | I | γ1 | 4,22 | R W Q E G N V F S C S V M H E A L H N H Y T Q K S L S L S P G |
| Daw | II | γ1 | 1,17 | M H E A L H N H Y T Q K S L S L S P G |
| Sa | | γ2 | | F S C S V M |
| Wan | | γ2 | | M H E A L H N H Y T Q K S L S L S P G |
| Zuc | III | γ3 | 11 | F S(C,S,V)M H E A L H N R F T Q K S L S L S P G |
| Mar | | γ3 | 11 | M H E(A,L,H,N)R/F T Q K/S L S(L,S,P,G) |
| Vil | | γ3 | 21 | M H E A L H N R Y T Q K S L S L S P G |
| Vin | III | γ4 | | R(W,Z,Z,G,B,V,F=S,C,S,V)M(H,Z,A,L=H,B,H,Y)T Q K S L S L S L G |
| Ger | | γ4 | | M H E A L H N H Y T Q K S L S L S L G |
| She | | γ4 | | M H(Z,A,L,H,B,H,Y,T,Q)K/S(L,S,L,S,L,G) |
| RABBIT | normal | γ | | E W Q R G D V F T C S V M H E A L H N H Y T Q K S I S R S P G |
| G.PIG | normal | γ2 | | S(Y,Z,A,G,B,V,T,T,C,S,V)M H E A L H N H V T Q K A I S R S P G |
| HORSE | normal | γ | | M H E A L H N H Y T Q K S V S K S P G |
| HORSE | normal | γT | | M H E A V E N H Y T Q K N V S H S P G |
| COW | normal | γ1 | | M H E A L H N H Y T Q K S T S K S A G |
| COW | normal | γ2 | | M H Z A L H B H Y M Q K S T S K S A G |
| HUMAN | | | | |
| Ou | II | μ | | E(W)N T G Q T Y T C V V A H E A L P B R V T E R T V D K S T G T K P L Y B V S L V M S D T A G T C Y |
| unnamed | | μ | | A G T C Y |
| unnamed | | α | | T C Y |
| MOUSE | | α | (unnamed) | (I,C)Y |

**Table 2–5.** Some representative C-region subclasses in various species. Except for the references cited, these data were taken from Grey (1969a).

| Class | Species | Subclasses | Sequence differences | Interchain disulfide differences | References |
|---|---|---|---|---|---|
| $C_\lambda$ | Man, chimp | $C_{\lambda arg}$,$C_{\lambda lys}$ | See Table 2–3B | | 1 |
| $C_\gamma$ | Man | $C_{\gamma 1}$,$C_{\gamma 2}$,$C_{\gamma 3}$,$C_{\gamma 4}$ | See Table 2–4B | + | |
| | Mouse | $C_{\gamma 1}$,$C_{\gamma 2a}$,$C_{\gamma 2b}$ | See Table 2–4B | + | |
| | Cow | $C_{\gamma 1}$,$C_{\gamma 2}$ | See Table 2–4B | | |
| | Horse | $C_{\gamma 2a}$,$C_{\gamma 2b}$,$C_{\gamma 2c}$,$C_{\gamma T}$ | See Table 2–4B | | |
| $C_\alpha$ | Man[a] | $C_{\alpha 1}$,$C_{\alpha 2}$ | | +[a] | 2 |

[a]Evidence for a further division of human $IgA_2$ proteins into two nonallelic "sub-subclasses" is cited by Grey et al. (1968). In both $IgA_1$ and $IgA_2$ myeloma proteins the L-H disulfide bond was said to be absent; later (Jerry, Kunkel, and Grey, 1970), absence of the L-H disulfide bond was said to be characteristic of one allelic variant (*Am* 1+) of $IgA_2$. The exact relation between the "sub-subclasses" and the allelic variants has not been explained.

REFERENCES:

1. Appella and Ein (1967); Ein (1968); Ein and Fahey (1967); Hood and Ein (1968a); Hood, Grant, and Sox (1970).
2. Grey et al. (1968); Jerry, Kunkel, and Grey (1970).

according to the class of its H chain: that is, IgG molecules have $\gamma$ chains and IgM have $\mu$ chains, for example. The major immunoglobulin classes that have been described in various vertebrate species are listed in Table 2-6, along with some of their diagnostic features.

The intrachain disulfide bonds of the immunoglobulins are remarkably constant (Fig. 2-4). Interchain disulfide bonds, on the other hand, differ depending upon the C-region classes and subclasses present (Fig. 2-5).

So far, the distinction between V and C regions has been made on the basis of homology. There is another criterion—variability—by which the same distinction can be made, and V regions defined by variability are almost exactly coextensive with those defined by homology. As can be seen in Tables 2-2, 2-3, and 2-4, sequences from several independently isolated immunoglobulin chains with the same C-region class and subclass are available, and (apart from minor variations, most of which correlate with allelic differences—see Appendix B) C regions of the same class and subclass do not vary, while their associated V regions vary extensively. These tables show the remarkable extent of mutually consistent sequence data for the C regions of different $\kappa$, $\lambda$, and H chains; by contrast, only

**Table 2-6.** Immunoglobulin classes in various vertebrate species. Summarized from the review articles cited below (references 1–4).

| Class | Species | Molecular weight of H chain (daltons) | | Polymeric form | Percent CHO | $C_H$ sequence known? | Other features | References |
|---|---|---|---|---|---|---|---|---|
| | | with CHO | without CHO | | | | | |
| IgM | Man | 65–70,000 | 60,000[b] | 5[c] | 7–13 | + | IgM may be associated with J chain | 5 |
| | Rabbit | 65–70,000 | 56–60,000 | 5[c] | 9.0 hexose, 3.3 hexosamine | | | |
| | Birds | 70,000 | | 5[c] | 2.6 hexose | | | |
| | Amphibians | 72,000 | | 6[c] | 10.8 | | | |
| | Bony fish | 70,000 | | 4[d] | 4.9–6.8 | | | 6–8 |
| | Elasmobranchs | 71,000 | | 1,5[e] | 3.7 hexose | | | |
| | Lamprey | 70–77,000 | | 1,5(?)[e] | | | | |
| IgG | Man | 50,000[b] | 48,600[b] | 1 | 4.8 | + | | |
| | Mouse | | | 1 | | + | | |
| | Rabbit | 53,000 | 49,000[b] | 1 | 1.4 hexose, 1.5 hexosamine | + | | |
| | Birds | 67,000 | | 1 | 2.2 hexose | | | |
| | Amphibians | 54,000 | | 1 | 2.1 | | | |
| IgA | Man $IgA_1$ | 56–58,000 | 50,000 | 1,2, . . . in serum | | + | Secretory dimer found with J chain and secretory piece in secretions. | 5,9 |
| | Man $IgA_2$ | 52–53,000 | 46,000 | 2 + J + secretory piece in secretions | | + | | |
| | Rabbit | 64,000 | | | 3.2 hexose, 3.2 hexosamine | | | |
| | Mouse | 51–53,000 | | 1,2, . . . in serum | | + | Also demonstrated in mouse colostrum, but quaternary structure unknown. | 10 |

| | | | | | | | | |
|---|---|---|---|---|---|---|---|---|
| IgE[a] | Man | 71–73,000 | 62–63,000 | 1 | 14 | | "Reaginic" antibody; | 11 |
| | Rabbit | | | 1 | | | rabbit and human IgE cross-react. | 12,13 |
| IgD | Man | 60,000 | 50,000 | 1 | 11.3–14 | + | | 14–17 |

[a]A review in preparation by Dorrington and Bennich is referred to in Dorrington and Tanford (1970).
[b]Accurate estimates based on the sequence and composition of peptides.
[c]In mammals, birds, and amphibians the monomeric and pentameric (or hexameric) immunoglobulins differ extensively in their H chains.
[d]Although Metzger (1970) quoted an unpublished report of monomeric immunoglobulin in bony fish, none of the other workers cited succeeded in demonstrating it.
[e]The H chain in the monomeric and polymeric forms of immunoglobulin in the elasmobranchs are indistinguishable by the criteria of molecular weight, carbohydrate content, peptide mapping, and antigenic determinants. There is no evidence for another class of monomer comparable to mammalian IgG.

REFERENCES:

*Reviews*

1. Grey (1969a). (General review of evolution of immunoglobulins).
2. Metzger (1970). (Review of IgM structure).
3. Tomasi and Beinenstock (1968). (Review of IgA structure).
4. Dorrington and Tanford (1970). (Critical discussion of physical structure, including H-chain molecular weight).

*Articles*

5. Halpern and Koshland (1970).
6. Acton et al. (1971).
7. Bradshaw, Clem, and Sigel (1969).
8. Trump (1970).
9. Mestecky, Zikan, and Butler (1971).
10. Grey, Shei, and Shalitin (1970).
11. Ishizaka and Ishizaka (1967).
12. Kindt and Todd (1969).
13. Zvaifler and Robinson (1969).
14. Henney et al. (1969).
15. Rowe, Dolder, and Welscher (1969).
16. Saha et al. (1970).
17. Spiegelberg, Prahl, and Grey (1970).

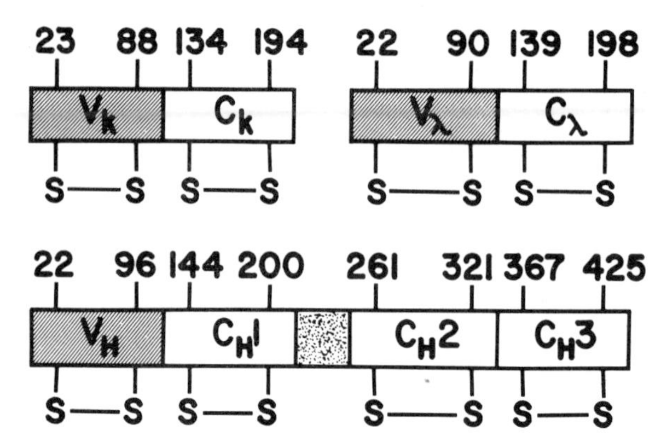


**Figure 2-4.** Intrachain disulfide bonds of the immunoglobulins. Intrachain disulfides usually can be experimentally differentiated from other disulfide bonds because they are only cleaved by sulfhydryl reagents if the chain is denatured (for example, by strong urea or guanidine solutions).

This bond has been demonstrated in human $\kappa$–Ag (Titani, Shinoda, and Putnam, 1969), Eu (Edelman et al., 1969), Ker, BJ, and Rad (Milstein, 1966c); human $\lambda$–Kern (Dayhoff, 1969), X (Milstein, 1966a), and Sh (Wikler et al., 1967); pig $\lambda$ (Dayhoff, 1969; Franek and Novotny, 1969); human $\gamma1$–Eu (Edelman et al., 1969), Daw (Piggot and Press, 1967), Cor (Press and Hogg, 1970), Dee (Dayhoff, 1969; Frangione, Milstein, and Pink, 1969), Car and Cra (Frangione, Milstein, and Pink, 1969); human $\gamma2$–Sa (Frangione, Milstein, and Pink, 1969); human $\gamma3$–Zuc, Kup, and Bru (Frangione, Milstein, and Pink, 1969); human $\gamma4$–Vin (Milstein, 1969b; Frangione, Milstein, and Pink, 1969); rabbit $\gamma$ (O'Donnell, Frangione, and Porter, 1970); and human $\mu$–Ou (Dayhoff, 1969; Shimizu et al., 1971).

Extra intrachain disulfide bonds have been found in human $\gamma1$ Daw (linking positions 33b and 100; Piggot and Press, 1967); rabbit $\gamma$ (linking position 131 or 132 to 220; O'Donnell, Frangione, and Porter, 1970); and rabbit $\kappa$ (positions unknown; Appella et al., 1970). An extra cysteine is found at position 33b of human $\gamma1$ Cor (Press and Hogg, 1970).

extremely rarely are two independently isolated chains identical in their V regions.

Antibody directed against any given determinant generally can be found associated with all possible classes, subclasses, and allelic forms of C regions; it is the V-region sequence variation, therefore, which must be responsible for the different specificities of antibodies. This conclusion is supported by the finding of hapten-specific amino acid compositional differences in $V_H$-region peptides from antibodies directed against different haptens (Koshland, 1967). Theories of antibody diversity, then, must not only explain the diversity itself, but also account for its restriction to one part of antibody polypeptide chains.

Allelic differences, called *allotypes*, have been described in several of the immunoglobulin chains, and genetic crosses, pedigree analysis, and popula-

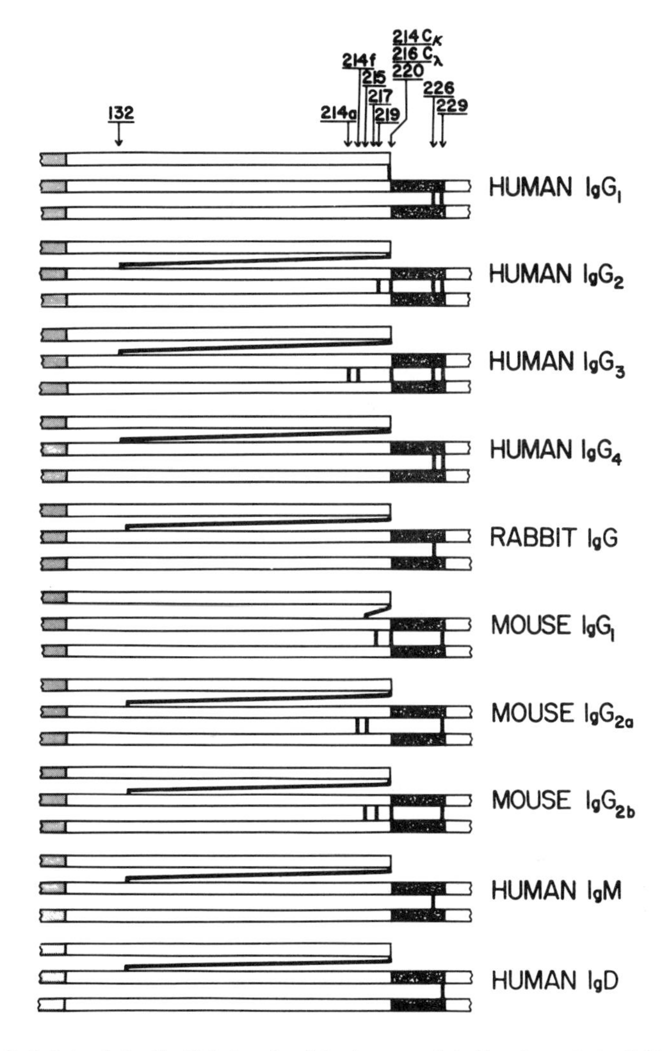


**Figure 2-5.** Interchain disulfide bonds of the immunoglobulins. Interchain disulfide bonds can be distinguished experimentally from intrachain bonds because they can be cleaved by sulfhydryl reagents under mild, nondenaturing conditions.

These bonds have been demonstrated in the following proteins: human $IgG_1$–Eu (Edelman et al., 1969), Daw (Steiner and Porter, 1967), Dee, Car, and Fie (Frangione, Milstein, and Pink, 1969); human $IgG_2$–Sa (Frangione, Milstein, and Pink, 1969); human $IgG_3$–Zuc (Frangione and Milstein, 1968), Kup and Bru (Frangione, Milstein, and Pink, 1969); human $IgG_4$–Vin (Frangione, Milstein, and Pink, 1969); rabbit IgG (O'Donnell, Frangione, and Porter, 1970); mouse $IgG_1$–MOPC 21 (Svasti and Milstein, 1970); mouse $IgG_{2a}$–MOPC 173 and MP 5563 (de Preval, Pink, and Milstein, 1970); mouse $IgG_{2b}$–MOPC 141 (de Preval, Pink, and Milstein, 1970); human IgM–Ou (Kohler et al., 1970b; Paul et al., 1971), Ale (Pink and Milstein, 1967); human IgD–Er (Perry and Milstein, 1970).

In mouse IgA and human $IgA_2$ of allotype *Am* 2+, L chains are disulfide bonded to one another rather than to H chains, while L chains are bonded to H chains in human $IgA_1$ and human $IgA_2$ of allotype *Am* 2– (Abel and Grey, 1968; Grey, 1969b; Grey et al., 1968; Jerry, Kunkel, and Grey, 1970).

tion studies have permitted the construction of genetic maps of the location of several of the immunoglobulin structural genes (see Appendix B). These maps for man, mouse, and rabbit will be shown in Fig. 9-2.

The amino acid sequence data on which the conclusions presented in this book are primarily based come from two sources, normal immunoglobulins and homogeneous proteins. Normal immunoglobulin chains, whether from unimmunized animals (Reisfeld, 1967) or from specific antibody preparations–see reviews by Haber (1968) and Haber et al. (1967)–are highly heterogeneous mixtures of sequences, most of whose differences are probably in their V regions. Sequence analysis of V regions from such mixtures is difficult; nonetheless, coherent sequences have been obtained from them. Since rare variant residues characteristic of only a few molecular species in the mixture are undoubtedly undetected in such studies, these sequences reflect the structure of the major prototypes of which the individual polypeptides present are variants. A sequence determined by Edman degradation for the first eighteen residues of normal human κ chains, for example, reflected the three prototype sequences which have been constructed independently by analysis of myeloma κ chains (Niall and Edman, 1967). These prototype structures determined from normal heterogeneous immunoglobulin chains will be called "average" sequences for short.

Although indirect conclusions about the origin of antibody variability can be drawn from average sequences, homogeneous proteins allow antibody variation to be directly examined and thus provide the most valuable evidence on it origin. Most such homogeneous immunoglobulins are produced by plasma-cell tumors (also called myelomas); with rare exceptions, each tumor excretes a unique, homogeneous product. By every available criterion (sequence, antigenic, and genetic analysis) these chains appear to be typical representatives of the corresponding chains in normal immunoglobulin. The myeloma tumors of mouse and man produce two types of products, both of which I shall call myeloma proteins: whole immunoglobulin molecules, consisting of both H and L chains; and the Bence-Jones proteins, which are free L chains (or L-chain dimers) excreted in the urine. In addition to myeloma proteins, homogeneous immunoglobulins can be obtained from normal animals by special immunization procedures.

This book will be concerned primarily with the evolution of V regions, since they contain the main clues concerning the origin and control of antibody diversity; the evolution of C regions is discussed briefly in Chapter 4. First, however, it will be necessary to explain the methods by which the evolutionary history of these proteins can be reconstructed.

# Chapter 3

## Reconstructing Protein Evolution

Many of the current ideas in theoretical immunology find their most direct, objective, and economical expression in simple statements about the evolution of antibodies. In this book I present a systematic survey of antibody evolution, as it can be inferred from amino acid sequence information, and discuss its implications for the central theoretical questions of the origin of antibody diversity and the selective control by antigen. A preliminary account of the earlier phases of this project has been published (Smith, Hood, and Fitch, 1971). This chapter will explain the methods used in reconstructing antibody evolution, Chapter 4 will present the results for C regions, and Chapter 6 those for V regions.

The reconstruction of the evolutionary history of a family of related proteins, in the absence of fossil proteins, must rely solely on the information residing in contemporary amino acid sequences. Most methods are based on the simple premise that of all the possible genealogies describing the evolution of a given set of proteins, the most likely to be correct is that which requires the fewest mutations. This will be called the principle of parsimony. The genealogies reported in this book have been reconstructed basically according to this principle, using two different methods. In the *method of minimum mutations*, all possible genealogies relating a given set of proteins are considered, the minimum number of mutations required at each nucleotide position being tallied for each genealogy. Because the number of possible genealogies increases rapidly as more sequences are included, this method, even with the aid of a computer, is limited to rather small numbers of proteins; I have not applied it to more than seven at one time. The *method of Fitch and Margoliash* (1967, 1968), on the other hand, does not examine every alternative genealogy, but rather chooses from among the possibilities those which might be said to be the most likely candidates. It is thus suitable for large numbers of sequences. Furthermore, the final choice among these candidates is not made directly on the basis of the number of mutations they require, but is based on a far more easily computed index of "badness"–the percentage of standard deviation (S.D.)–which the method seeks to minimize.

The first step in reconstruction by either method is alignment of the amino acid sequences, so that homologous residues occupy the same position in all the sequences. The amino acids found in a given position in properly aligned sequences all derive from the same position in their common ancestor. As can be seen in the alignments in Tables 2-2, 2-3, and 2-4, this often requires that gaps of one to several positions be inserted in some of the sequences: these presumably represent insertions or deletions that have occurred in evolution and introduce uncertainties in the alignment.

## The Method of Minimum Mutations

When the minimum mutation method is to be used, the amino acid sequences are translated into their corresponding mRNA nucleotide sequences. Each possible genealogy is then examined in turn, and the minimum number of mutations required at each nucleotide position is computed and recorded. Figure 3-1 illustrates this computation for a single hypothetical nucleotide position and defines the critical terms *excess mutations* and *informative positions*. It also shows how a *common ancestor* sequence can be constructed for a set of sequences which are assumed to have evolved according to a particular descent.

The translation of an amino acid sequence into its corresponding nucleotide sequence is generally straightforward. At the third position in the codon, where several nucleotides are possible, all of them are entered. When the amino acid is Arg, Ser, or Leu, however, there is some ambiguity which is resolved according to the rules in Table 3-1. The one-letter abbreviations for nucleotides are given in the legend to Fig. 3-2. Nucleotide positions will be given a decimal number: the number before the decimal point refers to the amino acid (codon) position, the number after the point (which can be 1, 2, or 3) refers to the nucleotide position within that codon, reading from the 5′ end of mRNA.

In Fig. 3-2 the method of minimum mutations is applied to the reconstruction of a hypothetical descent of four simple proteins. This figure also defines *dislocations*, which are discussed below.

The method of minimum mutations (as well as that of Fitch and Margoliash) finds a best topology, which (because it requires the fewest mutations) is considered to be the most likely reconstruction of the descent of the proteins considered. As explained in Fig. 3-2, however, the topology so chosen is compatible with several possible descents, depending

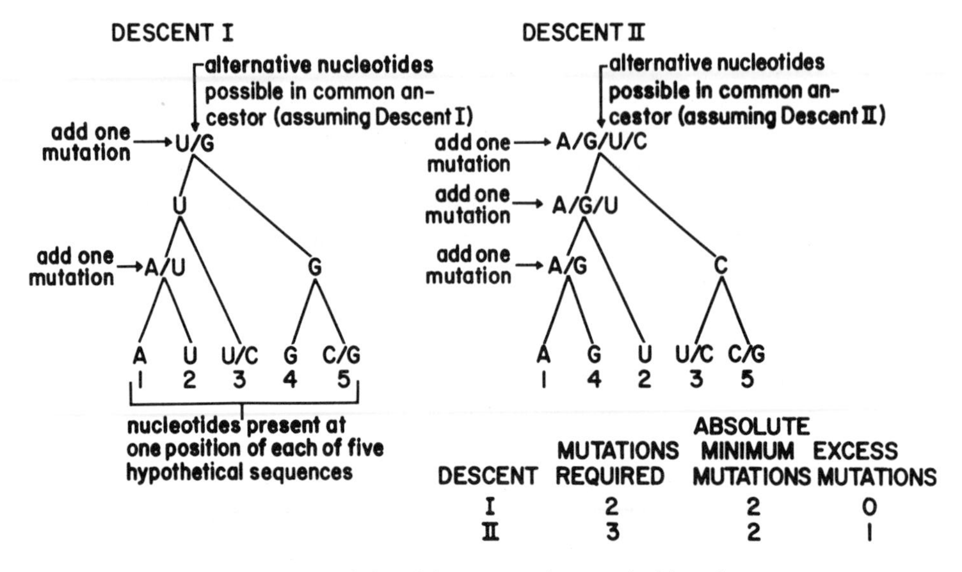


**Figure 3-1.** Computation of the minimum mutations required by a descent at a particular nucleotide position.

*Possible topologies*: For a given number of sequences, there are a certain number of topologically distinct ways by which they could have evolved. Each topology is in turn compatible with several descents, depending on which of its branches the common ancestor is thought to join, as illustrated in Fig. 3-2. Here the nucleotide (or alternative nucleotides) present at a particular position in each of five hypothetical sequences is shown, and computation of the minimum number of mutations required by two of the fifteen topologically distinct descents relating them is illustrated. To reconstruct a protein genealogy, the process here illustrated for a single nucleotide position is applied to all the nucleotide positions, often numbering in the hundreds.

*Minimum mutations required by a descent at one position*: We can compute the minimum mutations which must have occurred at this position for any of the possible topologies. (Any descent compatible with a given topology can be used; the same minimum mutations are required by all the other descents compatible with that topology as well.) Successively higher nodes are created in the descent, according to the following rules: (1) if any nucleotides are common to the two nodes being joined, these common nucleotides are entered at their junction node; (2) if no nucleotides are common to the two nodes being joined, *all* the nucleotides which occur in them are entered at their junction node, and one mutation is added to the tally of mutations required. When this process has been completed, the common ancestor node being thereby created, the mutations required are just the number of nodes at which the two lower nodes joined had no common nucleotides.

In the two descents shown in this figure, the nodes created according to these rules are entered, and those nodes at which a mutation could be deduced are indicated.

*Common ancestor*: The above process also created a common ancestor for the descent to which it was applied. The nucleotides which are entered at the highest node are those which might appear in the common ancestor at that position, if the sequences are assumed to have evolved according to that descent. The common ancestor, each of whose nucleotide positions is so defined, is the set of all nucleotide sequences from which the actual known sequences could have evolved according to that descent with the minimum number of mutations. Obviously, a common ancestor on where the common ancestor joins its branches. The actual descent cannot be chosen without either additional assumptions or additional information in the form of sequences more distantly related to the ones studied. In this book the actual descent usually is not as important as its topology, and the terms descent, genealogy, tree, and topology will often be used interchangeably. Reconstructed genealogies will be drawn in the from of descents, but it must be borne in mind that the placement of the common ancestor cannot be determined solely from the information and the premises used in the reconstruction.

It is clear that the deletion or insertion of a few amino acids is a type of heritable mutation analogous to simple nucleotide replacement. In the minimum mutation method the two types can be treated in the same way, and occasionally a tally of excess deletions and insertions (which will be grouped together under the term *gaps*) is kept separately from that of replacements, because it is not clear what relative weight the two types of mutations should be given. Otherwise these gaps are ignored. Almost always the genealogies which are most parsimonious of necleotide replacements are also those which are most parsimonious of gaps.

The data accumulated by the method of minimum mutations will be reported in several ways. For each set of sequences examined, the reconstructed genealogy (that is, the best topology) will be drawn. To give an

---

is defined only with reference to a particular descent and does not apply to other descents compatible with the same topology, since they place the common ancestor at a different position in the topology.

*Absolute minimum mutations at a nucleotide position*: For every nucleotide position there are one or more best topologies which require fewer mutations than the other topologies. In this case, Descent I is actually one of the best, while Descent II requires one more mutation. The mutations required by the best topologies for a given nucleotide position are called the *absolute minimum mutations.* Their number can easily be computed by subtracting 1 from the minimum number of different nucleotides which occur at that position; it thus ranges from 0 to 3. For a set of nucleotide sequences with many positions, the absolute minimum number of mutations is the sum of the absolute minimum mutations for each individual nucleotide position.

*Informative positions*: It often happens that all possible topologies are best for a given nucleotide position and require the same number of mutations. Such a position is called uninformative, since it does not aid in the reconstruction of a genealogy. The other positions, at which some topologies are better than others, are called *informative.*

*Excess mutations:* A more meaningful picture of the genealogical evidence residing in a given nucleotide position can be obtained if the absolute minimum mutations are subtracted from the mutations required by each topology, to give what will be called the *excess mutations* at that position. Uninformative positions then require zero excess mutations for all topologies. The table at the bottom of the figure gives the minimum mutations, absolute minimum mutations, and excess mutations for the two descents examined.

**Table 3-1.** Choosing between alternative codons for Ser, Arg, and Leu.

The computer program which calculates minimum mutations works on each nucleotide position independently of the others. This creates some ambiguity as to which nucleotide codons should be used for Ser, Arg, and Leu, for the choices possible at one position in these codons depend on the nucleotides chosen at another. In each of these cases, one of the two possible (not necessarily exclusive) sets of codons in which the ambiguity at one position does *not* depend on that at another is chosen by the following rules:

*Rule 1*: The absolute minimum mutations (see Fig. 3-1) for the three nucleotide positions should be minimized. Thus in Example 1, AGY is chosen for Ser because in that case the absolute minimum mutations is one less than if UCX were chosen.

*Rule 2*: If a choice remains after Rule 1 is satisfied, that codon is chosen which minimizes the genealogical informativeness of the three nucleotide positions. Thus in Example 2, CGX is chosen as the translation of Arg, because in that case there are no informative positions (see Fig. 3-2) in the codon.

*Example 1*

| Amino acids | Codons |
|---|---|
| Ser | U C X, or |
| | A G Y |
| Ala | G C X |
| Gly | G G X |
| Asn | A A Y |

| if Ser is: | Abs. min. is: | Total abs. min. mutns. | No. of inform. posns. |
|---|---|---|---|
| U C X | 2 2 0 | 4 | 0 |
| A G Y | 1 2 0 | 3 | 1 |

Choose AGY for Ser, since this choice minimizes total absolute minimum mutations.

*Example 2*

| Amino acids | Codons |
|---|---|
| Arg | C G X, or |
| | S G R |
| Asn | A A Y |
| Tyr | U A Y |
| Phe | U U Y |

| if Arg is: | Abs. min. is: | Total abs. min. mutns. | No. of inform. posns. |
|---|---|---|---|
| C G X | 2 2 0 | 4 | 0 |
| S G R | 1 2 1 | 4 | 1 |

Choose CGX for Arg, since this minimizes the informativeness of the position.

immediate idea of the degree of certainty of the genealogy so reconstructed—in other words, the degree of separation of its various nodes—mutations will be assigned to its branches as follows. Each topology consists of *terminal* branches (branches terminating in one of the actual sequences used in the reconstruction) and *connecting* branches (those connecting the nodes in the topology to one another). It is the connecting branches which determine the structure of the topology. Each such branch divides the sequences into two groups, and the number assigned to it will be the number of extra mutations (over the number required by the best topology) required by the most parsimonious alternative topology in which this grouping is not preserved. This number represents the minimum "cost" in extra mutations of obliterating that branch. If no parallel or back-mutations occur, this number will simply equal the minimum number of mutations which must be assigned to that branch in the best tree. Parallel or back-mutations, however, reduce the cost of rearranging the topology and therefore decrease the length of connecting branches below this number.

> This procedure underestimates the actual evidence for mutations occurring in the connecting branches. Suppose, for example, that we are relating five proteins, two of which have Adenine, two Guanine, and one Cytosine at a particular nucleotide position. There is clear evidence for a Guanine-to-Adenine mutation at this position in a connecting branch between the first and second pair of proteins, yet because the fifth protein can be joined without cost to any branch of the descent, all connecting branches will be assigned zero mutations at this position.

In addition to a drawing of the best topology, the distribution of total excess mutations required by all the alternative topologies will be shown, either as a bar graph or as a two-dimensional "scattergram" (Chapter 4). The actual structures of the "poor" genealogies will not be indicated. In a few cases, a table of the excess mutations required by each topology at each informative nucleotide or codon position will also be included; in these tables dislocations are in large type, and except for dislocations, zeros are not entered. The construction of such a table of excess mutations is illustrated in Fig. 3-2.

In many sets of sequences examined by the method of minimum mutations, some of the sequences included are *common ancestors*, constructed (as described in Fig. 3-1) from other proteins by *assuming* a particular descent for them. This assumed descent will be indicated by dotted lines in the best topology, and a brief justification for it will be given.

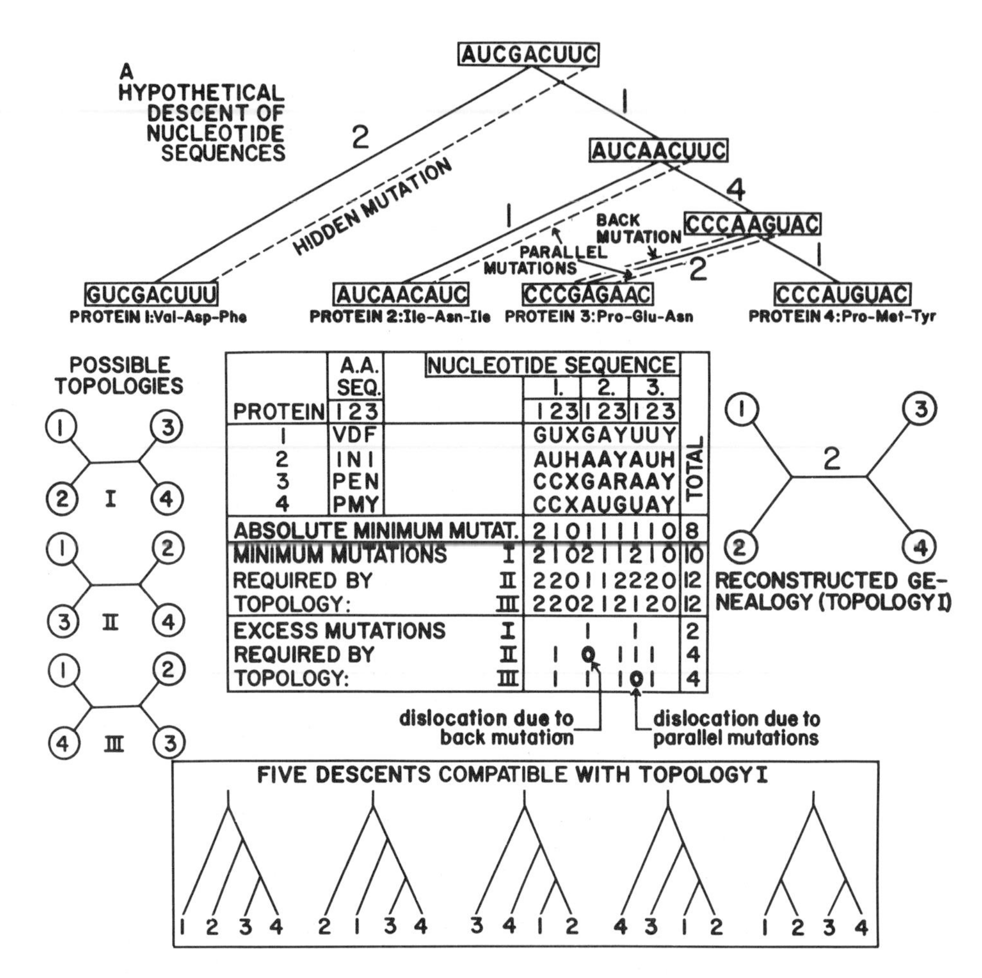


**Figure 3–2.** A hypothetical descent, and its reconstruction by the method of minimum mutations. Overall, Topology I, which conforms to the actual descent, requires two fewer mutations than the alternative topologies, II and III, as shown in the tables of minimum and excess mutations. It therefore represents the best reconstruction of the descent of these four hypothetical proteins.

*Dislocations*: Topology I, although the best overall, is not the best at all positions. Because of a back-mutation which occurred in Protein 3, Topology II is better than I at nucleotide position 2.1. And because of parallel mutations in Proteins 2 and 3, Topology III is better than I at nucleotide position 3.1. In cases like these, where an alternative topology requires one (or more) fewer mutations at a given nucleotide position than the overall best topology, we say that one (or more) *dislocation* occurs in that alternative topology at that nucleotide position. In tables of excess mutations, dislocations are signaled by printing those entries in large type. Dislocations are expected to occur occasionally in any descent, as a result of parallel- and back-mutations. They can also occur by recombination (see Fig. 6–4).

*Assignment of mutations to connecting branches of a reconstructed topology*:

**The Method of Fitch and Margoliash**

The minimum mutation method described above is attractive in that it applies the principle of parsimony directly to the choice of the best topology, but it is only appropriate when small numbers of sequences are being examined. The number of distinct topologies ($t$), which is related to the number of sequences ($s$) by the expression $t = 3 \times 5 \times 7 \times \ldots \times (2s - 5)$, rises to impractically large numbers when $s$ exceeds 7.

The method described by Fitch and Margoliash (1967, 1968) circumvents this problem by eliminating from consideration the vast majority of the possible topologies which (by a certain criterion) are deemed unlikely to represent the true descent. This criterion of preliminary acceptability of a topology is, loosely stated, that sequences which are relatively similar should for the most part be more closely related in an "acceptable" genealogy than sequences which are relatively dissimilar. The degree of acceptable violation of this rule can be set in advance and determines the number of alternative topologies which will be considered. The actual procedure is described in the above references.

It is easy to construct a descent for which the most parsimonious of all the possible topologies grossly violates the acceptability criterion of this method; it is not necessarily true, that is, that closely related sequences in the most parsimonious tree are also relatively similar. This method, therefore, makes a subsidiary assumption, namely, that all the lines of descent leading to the proteins accumulated approximately the same number of mutations; and its ability to handle large numbers of sequences is conse-

---

Two mutations were assigned to the connecting branch of the best topology (I), because the best alternative topologies (II and III) in which this grouping is not preserved require this number of extra mutations.

A mutation which occurred at position 3.3 of sequence 1 in the actual descent is not apparent in the deduced nucleotide sequences (which are inferred from the amino acid sequences) because of the degeneracy of the genetic code.

The reconstructed genealogy is compatible with the five descents shown, depending upon which branch the common ancestor joins. The first of these reflects the actual descent in this example. Choosing between the several descents compatible with a reconstructed genealogy requires additional assumptions or information.

*Single-letter amino acid code*: A=Ala, B=Asx, C=Cys, D=Asp, E=Glu, F=Phe, G=Gly, H=His, I=Ilu, K=Lys, L=Leu, M=Met, N=Asn, P=Pro, Q=Gln, R=Arg, S=Ser, T=Thr, V=Val, W=Trp, Y=Tyr, Z=Glx, ?=unknown.

*Single-letter nucleotide code*: A=Adenine; G-Guanine; U=Uracil; C=Cytosine;R=A or G; Y=U or C; S=A or C; T=A or U; W=U or G; Z=C or G; B=C, G, or U; D=A, G, or U; H=A, C, or U; V=A, C, or G; X=unknown.

quently bought at the price of introducing an additional degree of uncertainty.

The final criterion of "badness"—percent standard deviation (S.D.)—which is used to make the final choice between these acceptable candidates is defined and illustrated in Fig. 3-3, which accurately reconstructs the same hypothetical descent that was reconstructed in Fig. 3-2 by the method of minimum mutations. Percent S.D. is not a direct application of the principle of parsimony, but it has been shown that in the absence of parallel and back-mutations (which in any case are assumed to be relatively rare), the tree with the lowest percent S.D. is also the most parsimonious of mutations. Percent S.D. is much easier to compute than the minimum number of mutations and is therefore an appropriate approximation when large numbers of sequences are examined.

The program used for reconstruction by the method of Fitch and Margoliash could accommodate only a limited number of gaps; therefore, except in a few cases, positions at which any of the sequences have gaps or are unknown were ignored in the reconstruction. The few remaining gaps were treated as described in the above references. The data accumulated by this method will not be reported in detail. Only the best topology found will be shown, with branch lengths defined as described in Fig. 3-3.

The overall ability of the method to reconstruct evolutionary history accurately, given quite simple and plausible assumptions about the process by which mutations accumulated in evolution, has been demonstrated (Fitch and Margoliash, 1968). It is generally quite able to delineate all but the closest divergences, and in this book it is used in preliminary evolutionary surveys to define major, obvious evolutionary groups. The details of the divergences within and among those groups are then reconstructed by the method of minimum mutations.

## Limitations of Reconstruction Methods

The reconstruction of the actual family tree from an examination of the pattern of mutations in contemporary sequences is subject to two limitations: this pattern can be uninformative or misinformative about the actual evolution of the proteins.

A position is uninformative about the order in which a set of divergences occurred if all possible arrangements of those divergences require the same number of mutations at that position (see Fig. 3-1). Such a position arises when a connecting branch (a branch which does not terminate in an actual

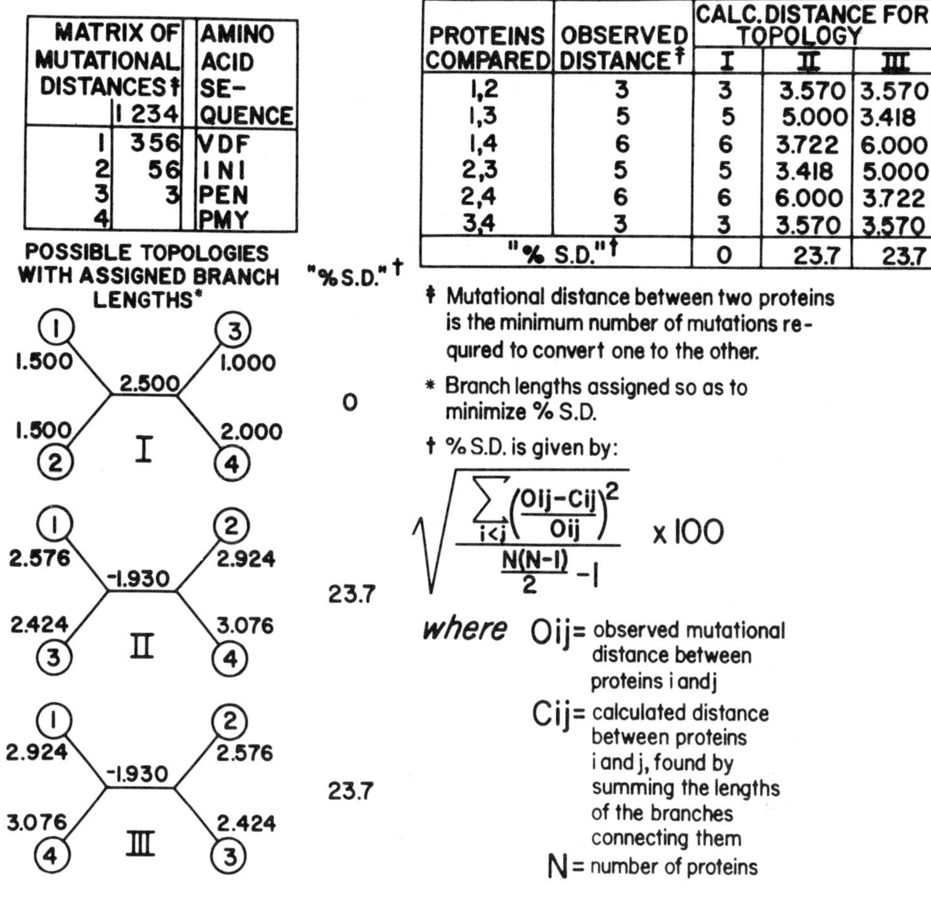

| MATRIX OF MUTATIONAL DISTANCES† | 1 | 2 | 3 | 4 | AMINO ACID SEQUENCE |
|---|---|---|---|---|---|
| 1 | | 3 | 5 | 6 | VDF |
| 2 | | | 5 | 6 | INI |
| 3 | | | | 3 | PEN |
| 4 | | | | | PMY |

| PROTEINS COMPARED | OBSERVED DISTANCE† | CALC. DISTANCE FOR TOPOLOGY I | II | III |
|---|---|---|---|---|
| 1,2 | 3 | 3 | 3.570 | 3.570 |
| 1,3 | 5 | 5 | 5.000 | 3.418 |
| 1,4 | 6 | 6 | 3.722 | 6.000 |
| 2,3 | 5 | 5 | 3.418 | 5.000 |
| 2,4 | 6 | 6 | 6.000 | 3.722 |
| 3,4 | 3 | 3 | 3.570 | 3.570 |
| "% S.D."† | | 0 | 23.7 | 23.7 |



$$\sqrt{\frac{\sum_{i<j}\left(\frac{O_{ij}-C_{ij}}{O_{ij}}\right)^2}{\frac{N(N-1)}{2}-1}} \times 100$$

**Figure 3-3.** Reconstruction of the descent in Fig. 3-2 by the method of Fitch and Margoliash. The best topology, I, which conforms to the actual descent, was shown in Fig. 3-2 to be the best topology also by the criterion of minimum mutations. Topologies II and III, which are poor in having a high percent standard deviation, are also anomalous in that a relatively large negative number must be assigned to their middle branch in order to minimize the percent standard deviation.

contemporary sequence) in the actual family tree accumulates no mutations, and it must often happen that such a branch accumulates no mutations at any position. Hence the divergences in a reconstructed genealogy are frequently not entirely ordered, and in these cases it is drawn with multiple nodes, from which more than two branches descend. I shall call a genealogy *highly ordered* if most of its divergences can be ordered with a high degree of confidence, and *unordered* if few of its divergences can be so ordered.

More fundamental than the problem of unordered divergences, which are an expected accident of the process of evolution, is the possibility that the principle of parsimony itself can lead to incorrect genealogical inferences. Parallel and back-mutations can result in a mutation pattern which points to incorrect conclusions. For example, in the hypothetical descent in Fig. 3-2, one pair of parallel mutations and one back-mutation occur; although Topology I, which accurately reflects the true descent in this example, is the overall best genealogy, both by the method of minimum mutation (Fig. 3-2) and by the method of Fitch and Margoliash (Fig. 3-3), it is not the best in every position. In position 3.1, Topology II requires one less mutation because of the parallel mutations in proteins 2 and 3, which make these proteins appear in this position to be related to each other. And at position 2.1 Topology III appears to be best because a mutation in protein 3 reversed one that had occurred in the common ancestor of proteins 2, 3, and 4. In the table of excess mutations in this figure, tallies for the alternative topologies (II and III) which are less than that at the same position for the overall best topology (I) are called *dislocations* and are printed in large type to call attention to possible points of uncertainty. This practice will be followed in all excess mutation tables in this book.

The correct reconstruction of the genealogy, therefore, depends in some way on the ratio of mutations which reflect the true history of the sequences to parallel and back-mutations which contradict their testimony. In the reconstruction in Fig. 3-2, for example, Topology I is still better than its alternatives despite the two dislocations.

It is clear that parallel and back-mutations do occur in nature. Even if mutations accumulated entirely at random, a certain number would be expected by chance. Furthermore, parallel mutations might be expected to occur even more frequently than expected by chance, since if a mutation has been accepted in one evolutionary line it is much more likely to be acceptable in another than randomly chosen mutations, most of which are presumably deleterious.

Although parallel and back-mutations do occur, it seems unlikely that they would significantly alter genealogical conclusions. In order to do so, they would not only have to occur at a high frequency, they would also have to be coordinated. For example, in Fig. 3-2, the parallel and back-mutations did *not* conspire together, and consequently the topology falsely signaled by the parallel mutations was not the same as that favored by the back-mutation. In the absence of coordinated parallel and back-mutations, then, the incorrect topological alternatives favored by one parallel or back-mutation should not in general be the same as those favored by another. This is reflected in the tables of excess mutations: the dislocations are more or less randomly scattered among the other topologies. Because the false evidence is thus scattered and contradictory, consistent evidence for the true descent will stand out clearly in the mutation pattern, and we will not be misled in our reconstruction.

As will be discussed in Chapter 6, recombination can also result in dislocations; in principle, this case can be distinguished from that of parallel and back-mutations.

Clearly, a certain amount of circumspection is called for in drawing conclusions by the parsimony principle. One cannot, for example, confidently choose between a pair of genealogies which differ by one excess mutation, since such a slight edge may easily be attributed to parallel or back-mutations. In the scattergrams or bar graphs of the distribution of excess mutations reported in this book, I have indicated the values corresponding to close-contending topologies, between which (in my estimation) a choice cannot be confidently made. Any of these close contenders can be taken as an acceptable reconstruction of the descent. In the diagrams of the genealogies reconstructed by the method of minimum mutations, dashed circles enclose the divergences which are rearranged in the close-contending topologies.

# Chapter 4

## The Evolution of C Regions

The immunoglobulin C regions within each of the three major families have evolved separately from the corresponding V regions and therefore do not bear directly on the detailed origin of variation in the latter. Nevertheless, in elucidation of the overall evolution of the various functions of the immune apparatus, C-region evolution will be prima facie evidence, since these regions are determinants of the nonspecific "effector" functions which the various types of antibody carry out regardless of specificity (Edelman et al., 1969). On the basis of a genealogical analysis of C-region amino acid sequences, and other less direct evidence, a general scheme for the descent of C regions can be proposed. In all but a few respects, it conforms to that developed by Grey (1969a) in a recent review.

The starting point of this analysis is the hypothesis of Hill and his associates (1966), amply confirmed by Edelman and his co-workers (1969), that all C regions are composed of linearly connected homology units, each about 110 residues in length, and all deriving from a single common ancestor. The homology (as measured by the statistical test of Fitch, 1966) between these units is much too great to admit a reasonable doubt of this premise. Furthermore, all the homology units so far studied share with one another (and with V regions) a common intrachain disulfide bond involving homologous cysteine residues (see Fig. 2-4). As diagrammed schematically in Fig. 2-1, a $C_L$ region consists of one such homology unit, while the $C_\gamma$ regions consist of three. Still in question is the number in the other $C_H$ regions ($C_\mu$, $C_\alpha$, $C_\delta$, $C_\epsilon$, and nonmammalian $C_H$).

Table 4-1 aligns those homology units for which a great deal of sequence information is available (or which are otherwise important for this discussion). The alignment is, with a few minor exceptions, the same as that proposed by Edelman et al. (1969), and the position numbers in this chapter refer to Table 4-1, not to the numbering system used for Tables 2-2 to 2-4, unless otherwise indicated.

Figure 4-1 is a genealogy reconstructed by the method of Fitch and Margoliash for a number of complete C-region homology units. It will be seen that the $C_L$ regions, and the three homology units $C_H1$, $C_H2$, and $C_H3$ of the $C_\gamma$ regions, form four separate families in this descent. This finding, taken in conjunction with the fact that the relations between human $C_{\gamma 1}$ and $C_{\gamma 4}$ and rabbit $C_{\gamma}$ are the same in both $C_H2$ and $C_H3$, suggest that a single series of fused duplications led to a single $C_H$ ancestor from which all $C_{\gamma}$ regions derived. I shall assume, by extension, that all other $C_H$ regions derived from the same common $C_H$ ancestor as well.

These assumptions were used in a number of analyses by the method of mimimum mutations, the results of which are summarized in Fig. 4–2. The best tree found in each set is diagrammed, and except for the genealogy in Fig. 4–2D (in which all possible topologies were not examined), mutations are assigned to the connecting branches as described in Chapter 3. Next to the best descent is some form of distribution of the number of excess mutations required by all the alternative genealogies. For Figs. 4–2A, 4–2B, and 4–2E this distribution is given as a bar graph, while for Fig. 4–2C it takes the form of a scattergram (a form which will be used extensively in the examination of V-region evolution). In this scattergram the excess mutations required by each genealogy in the first half of the sequences is plotted against the number of mutations required by the same topology for the other half; the number printed in the *i*th row of the *j*th column is thus the number of different topologies which required *i* excess mutations in the first half and *j* excess mutations in the second. The purpose of this form of presentation is to detect recombination in V regions, as will be discussed in Chapter 6. For the present discussion it can be regarded as merely another way of displaying the distribution of excess mutations; in addition, it provides a check on our conclusion, for it tells us to what extent the pattern of mutations in the first half and that in the second support the same topology. For this scattergram, as for every other reported in this book, the best or one of the best topologies for the first half is also the best or one of the best for the second, well within the uncertainties of the reconstruction. (I am indebted to E. H. Peters for writing the program which constructed these scattergrams.) In the bar graphs and scattergrams, I have indicated those topologies on the far left or in the lower left-hand corner which in my opinion include all the genealogies that cannot be considered definitely "worse" than the overall best, and the nodes rearranged in these close-contending alternatives are circled in dashed lines.

## $C_L$ Evolution

Few of the conclusions from this analysis are very surprising. Figure 4–2A shows that $C_{\kappa}$ regions from different species are more closely related to one another than to $C_{\lambda}$ regions, and vice versa. This finding implies that

```
                                        1                   2                   3                   4                   5                   6             6
                 (SUB)
UNIT SPECIES     CLASS 1 2 3 4 5 6 7 8 9 0 1 2 3 4 5 6 7 8 9 0 1 2 3 4 5 6 7 8 9 0 1 2 3 4 5 6 7 8 9 0 1 2 3 4 5 6 7 8 9 0 1 2 3 4 5 6 7 8 9 0 1 2 3 4 5 6 7
CL   Human       κ     - T V A A P S V F I F P P S D E Q - - L K S G T A S V V C L L N N F Y P R E A K V - - Q W K V D N A L Q S G N S Q E S V T E Q D S K D
     Mouse       κ     - A B A A P T V S I F P P S S E Q - - L T G G S A S V V C F L N N F Y P K D I N V - - K W K I D G S E R Q B G V L B S B T Z W D S K D
     Pig         κ                                                                                             W K V D G V V Q S S G I L D S V T E Q D S K D
     Human       λ     Q P K A A P S V T L F P P S S E E - - L Q A N K A T L V C L I S D F Y P G A V T V - - A W K A D S S P V K A G V E T T T P S K Q S - N
     Mouse       λ     Q P K S S P S V T L F P P S S E E - - L T E N K A T L V C T I T D F Y P G V V T V - - D W K V D G T P V T Q G M E T T E P S K Q S - N
CH1  Human       γ1    A S T K G P S V F P L A P S S K S - - T S G G T A A L G C L V K D Y F P E P V T V - - S W N S - G A L T S G - V H T F P A V L Q S - S
     Human       γ2                      P L A P C S R S - - T S E S T A A L G C L
     Human       γ3                      P L A P C S R S - - T S G G A A A L G C L
     Human       γ4    A S T K G P S V F P L A P C S R S - - T S E S T P A L G C L V K
     Rabbit      γ     S G T K A P S V F P L A P C C G D - - T P S S T V T L G C L V K G Y L P E P V T V - - T W N S - G T L T D G - V R T F P S V R Q S - S
     Guinea P.   γ2        T T A P S V F P L A V C S G B - - T A M S T L S M G C L V K G Y F P Z P V T V - - K
     Human       μ     G S A S A P T L F P L V S C E N S - -(B,P,S,S,T)V A V G C L A Z D F L P D S I T F - - S W K Y(B,B,S,B,K)I S S T R G F P S V L R - - G
CH2  Human       γ1    - L L G G P S V F L F P P K P K D T L M I S R T P E V T C V V V D V S H E D P Q V K F N W Y V D G - V Q V H N A K T K P R E Q Q Y - B
     Human       γ2    - A V S F                               I S R T P E V T C V
     Human       γ3    - L L G G P S V F L F P P K P K D T L M I S R T P E V T C V V V D
     Human       γ4    - F L G G P S V F L F P P K P K D T L M I S R T P E V T C V V V D V S Q E D P Z V Z F N W Y V D G - V E V H N A K T K P R E E Q F - B
     Rabbit      γ     - L L G G P S V F I F K P P P K D T L M I S R T P E V T C V V V D V S E D D P E(V,Z)F T W Y I B B - Z Z V R T A R P P L R E Q Q F - D
     Guinea P.   γ2    - B L G G P S V F I F P P K P K D T L M
     Human       μ     - Q D T A I R V F A I P P S F A S I F L T K S T K L
CH3  Human       γ1    - Q P R E P Q V Y T L P P S R E E - - M T K N Q V S L T C L V K G F Y P S D I A V - - E W E S N D - G E P E N Y K T T P P V L D S - D
     Human       γ2                                                          T C L
     Human       γ3                                                        L T C L
     Human       γ4    - Q P R E P Q V Y T L P P S Q E E - - M T K N Q V S L T C L V K G F Y P S D I A V - - E W Z S B B - G Z P Z B Y K T T P P V L D S - D
     Rabbit      γ     - E P L E P K V Y T M G P P R E Q - - L S S R S V S L T C M I D G F Y P S D I S V - - G W E K D G - K A E D D Y K T T P A V L D S - D
```

| UNIT | SPECIES | (SUB) CLASS | Sequence position: 78–89, 8a, 90–106, 6a–6f, 107–127 (7 … 8 … 9 … 10 … 11 … 12 … 12) |
|---|---|---|---|
| CL | Human | κ | S T Y S L S S T L T L S K A D Y E K H K L - Y A C E V T H Q G L S S P V - T K S - - - - - - F - - N R G E C |
| | Mouse | κ | S T Y S M S S T L T L T K B Z Y Z R H B S - Y T C Z A T H K T S T S P I - V K S - - - - - - F - - N R N E C |
| | Pig | κ | S T Y S L S L T L S L P T S Q Y L S H N S - Y S C E V T H K B Z C Z A |
| | Human | λ | N K Y A A S S Y L S L T P E Q W K S H K S - Y S C Q V T H E G S T - - V - E K T - - - - - - V - - A P T E C S |
| | Mouse | λ | N K Y M A S S Y L T L T R A W E R S H S Y - S S C Q V T H Z G H T - - V - Q K S - - - - - - L - - S R A D C S |
| CH1 | Human | γ1 | G L Y S L S S V V T V P S S S L G T Q - T - Y I C N V N H K P S N T K V - D K R - - - - - - V - - E P K S C D K T H T C P P C P A P E |
| | Human | γ2 | Y T C B V B H K P S B T K V - B K T - - - - - - V - - Z R K C C V Z - - - C P P C P A P G |
| | Human | γ3 | Y T C B V B H K P S K T P L - G D T(C,T,P,H,R)C P - - E P K S C D T P P P C P R C P A P E |
| | Human | γ4 | T - Y T C N V D H K P S N T K V - D K R - - - - - - V - - E S K - - - Y G P P C P P C P A S E |
| | Rabbit | γ | G L Y S V P S T V S V S Z P - - - - - - - P - S T C B V A H - A T B T K V - D K T - - - - - - V - - A P S T C S K - P T C P P - - - P E |
| | Guinea P. | γ2 | V T V P S S - - - - - - - E - K T C N V A H P A S T S K V - D K T - - - - - - V - - E P I R T P Z P(B,S)C P P - - - P E |
| | Human | μ | G K Y A A(Z,S,T)V L L P S K D V M Q G(H,B,T,Z)V C K V - D H R - - - - - - G - - L T F Z Z B A S S M C - - - - V P D |
| CH2 | Human | γ1 | S T Y R V V S V L T V L H Q N W L D G K E - Y K C K V S N K A L P A P I - E K T - - - - - - I S K A K G |
| | Human | γ2 | T K C K V S N K G L P A P I - E |
| | Human | γ3 | Y K C K V S N K A L P A P I - E |
| | Human | γ4 | S T Y R V V S V L T V L H Z B W L D G K E - Y K C K V S N K G L P S S I - E K T - - - - - - I S K A K G |
| | Rabbit | γ | S T I R V V S T L P I A H E D W L R G K E - F K C K V H D K A L P A P I - E K T - - - - - - I S K A R G |
| CH3 | Human | γ1 | G S F F L Y S K L T V D K S R W Q E G N V - F S C S V M H E A L H N H Y T Q K S - - - - - - L S L S P G |
| | Human | γ2 | F S C S V M H E A L H N H Y T Q K S - - - - - - L S L S P G |
| | Human | γ3 | F S C S V M H E A L H N R F T Q K S - - - - - - L S L S P G |
| | Human | γ4 | G S F F L Y S R L T V D K S R W Z Z G B V - F S C S V M H E A L H N H Y T Q K S - - - - - - L S L S L G |
| | Rabbit | γ | G S W F L Y S K L S V P T S E W Q R G D V - F T C S V M H E A L H N H Y T Q K S - - - - - - I S R S P G |
| | Guinea P. | γ2 | S Y Z A G B V - T T C S V M H E A L H N H V T Q K A - - - - - - I S R S P G |
| CH4? | Human | μ | G R Y F A H S I L T V S E E E(W)N T G Q T - Y T C V V A H E A L P B R V T E R T - - - - - - V D K S T G TKPLYBVSLVMSDTAGTCY |

**Table 4-1.** Alignment of C-region homology units.

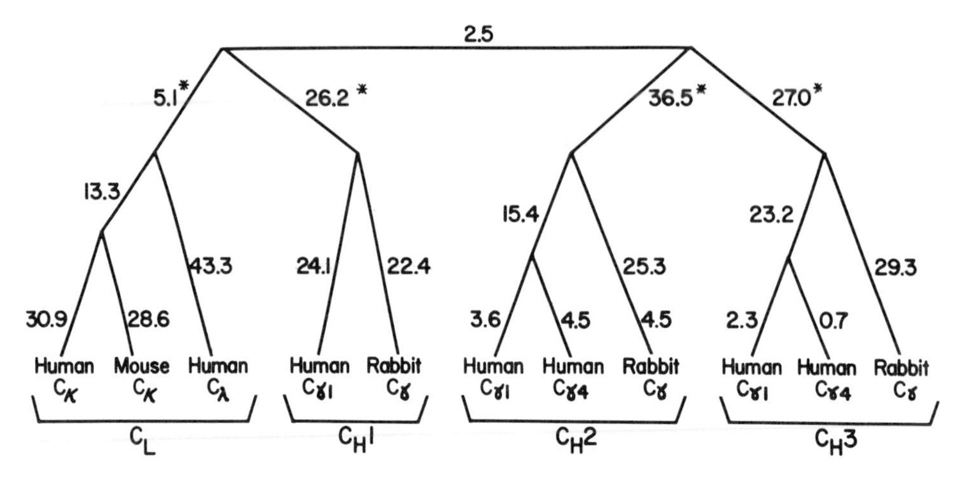


**Figure 4–1.** Best genealogy found in a preliminary survey of C-region evolution by the method of Fitch and Margoliash. The branches marked with an asterisk will be assumed to be correct in the subsequent analysis by the method of minimum mutations.

the $C_\kappa$ and $C_\lambda$ genes both were present in the common phylogenetic ancestor of mouse and man, a conclusion strongly supported by the finding of $C_\kappa$- and $C_\lambda$-like peptides in a wide variety of mammals (see Tables 2–2B and 2–3B). The order of divergence of these three $C_\kappa$ regions could not be determined. $V_\lambda$-, $C_\kappa$-, and $C_\lambda$-like peptides have all been described in the chicken (see Tables 2–2B and 2–3), but it is not established that they are not the various regions of a single class of $C_L$ in this species derived from the common ancestor of both mammalian $C_\kappa$ and $C_\lambda$. Thus we cannot yet tell whether the duplication leading to the two families occurred before or after the mammalian and bird phylogenetic lines diverged, but it certainly occurred before the mammals diverged from one another.

A very recent duplication of the $C_\lambda$ gene has occurred in the hominid line, leading to the $C_{\lambda arg}$ and $C_{\lambda lys}$ subclasses, which differ by a single nucleotide replacement at position 194; both subclasses are found in chimpanzees and man, but only one in monkeys and baboons (Tables 2-3B and 2-5). Gibson, Levanon, and Smithies (1971) and Hess and Hilschmann (1970) have presented evidence that a further duplication of the $C_{\lambda arg}$ gene in man is responsible for the Ser-Gly interchange observed at position 157 of some of these proteins.

## $C_H$ Evolution

Figure 4–2B, taken in conjunction with Fig. 4–1 (which places the point of earliest time in the topology), shows that the human $C_\gamma$ subclasses diverged from one another much more recently than the human, rabbit, and guinea pig $C_\gamma$ regions diverged. Thus the human subclasses certainly do not correspond to those of distantly related mammals such as the mouse, rabbit, or guinea pig, and any functional analogy between human subclasses and those of such species must be attributed to convergent evolution. This conclusion is supported by a few partial sequences from the three mouse $C_\gamma$ subclasses, which show no close relationship to those from man (see Table 2–4B). The same point is made by the genealogy in Fig. 4–2D, which suggests that the subclasses of the cow are more related to one another than to other $C_\gamma$ regions; this topology, however, rests on very limited sequence information. Taken together, these analyses demonstrate that the $C_\gamma$ genes have undergone several independent duplications in various mammalian lines.

Figure 4–2C summarizes an attempt to determine the relation of the $C_H1$ homology unit of human $C_\mu$ to $C_\gamma$ and $C_L$ regions. This sequence was, as expected, found to be definitely more related to the corresponding homology units of $C_\gamma$ than to other homology units. However, the relationships *within* the $C_H1$ units were surprising, in that (within the uncertainties of the method) the $C_\gamma$ regions of guinea pig, rabbit, and man were not found to be significantly more related to one another than to human $C_\mu$. The marked differences between $C_\mu$ and the various $C_\gamma$ regions thus seem to be attributable less to a distant relationship than to an extraordinarily large accumulation of mutations in this portion of $C_\mu$. Despite this finding, a large body of indirect evidence to be discussed presently suggests that $C_\mu$ and $C_\gamma$ diverged before the mammals diverged from the amphibians. More sequence data will be required to reconcile this apparent contradiction. Particularly helpful would be a $C_H$ sequence from some elasmobranch species, since they are presumed on the basis of the indirect evidence cited below to derive from a common ancestor of mammalian $C_H$ regions of all types.

Neither of the analyses in Figs. 4–2C and 4–2E reveals the order of the divergences among the $C_H1$, $C_H2$, $C_H3$, and $C_L$ homology units; nor do they reveal any significant evidence for recombination among these four ancestors. Similarly, the table of excess mutations (Table 4–2) for the

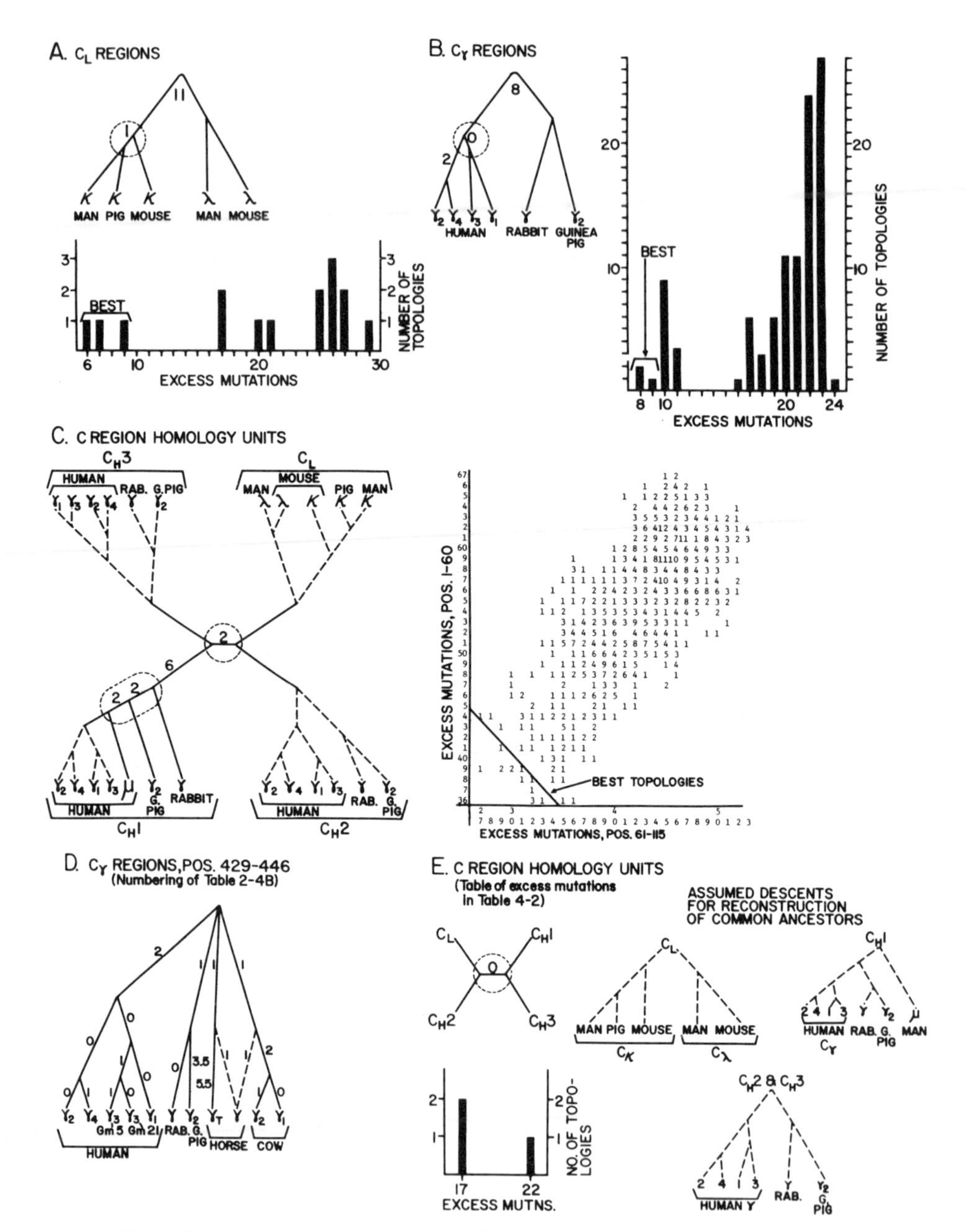


**Figure 4-2.** C-region evolution, method of minimum mutations. Assumed descents* are indicated by dashed lines.

**Justification for assumed descents*: In Figs. 4-2C and 4-2E, the human $C_\gamma$'s were assumed to have descended as shown and to be more closely related to one another than to other $C_H$ regions, on the basis of 4-2B and of Fig. 4-1 (which places their common ancestor with rabbit $C_\gamma$). Rabbit and guinea pig $C_\gamma$'s were assumed to be more closely related to one another than to other $C_H$ regions in 4-2C and 4-2E, but if this assumption were wrong there would have been no essential difference in the analysis in Fig. 4-2E shows a scattered pattern of dislocations, more reminiscent of parallel and back-mutations (Fig. 3-2) than of simple recombination (Fig. 6-4). Recombination might be expected if the ancestral $C_H$ gene was lengthened by a series of homologous but unequal crossings over. The failure to find this by no means indicates that such a recombination did not occur, however. It seems likely that the very large number of mutations which the $C_L$, $C_H1$, $C_H2$, and $C_H3$ regions have undergone since their divergence have gradually covered up the evidence for their evolutionary relation with subsequent superimposed mutations, leading to the very high level of dislocations actually observed.

A further discussion of $C_H$ evolution rests on indirect evidence (summarized in Table 2-6) concerning the presence or absence of various classes

```
       EXCESS MUTNS. AT NUCLEOTIDES
                                111
        111112223445556668888999001
TOP.   3401567568707572381245899171
 NO.   1222111321121221221132121221 2  TOTAL
   1   0011011110101001111111000 1101   17
BEST   11  111 1111 11    11111  1      17
   3   1111100101011111111001111111     22
```

Table 4-2. Excess mutations for Fig. 4-2E (C-region homology units).

of immunoglobulin in a variety of mammals, and in several nonmammalian vertebrates which have diverged from the mammals at successively earlier epochs: amphibia, bony fish, elasmobranchs (shark-like fish), and cyclostomes (lampreys, hagfish). It should be emphasized at the outset that this evolutionary scheme is built upon only the grossest characterization of the

---

results. The descent assumed for human, pig, and mouse $C_\kappa$'s and human and mouse $C_\lambda$'s in 4-2C and 4-2E was based on the results in 4-2A and on the evidence cited in the text that the $C_\kappa$-$C_\lambda$ divergence preceded the mammalian phylogenetic divergences.

The genealogy in 4-2D represents the two most parsimonious topologies (which differ only in the placement of horse $C_\gamma$) for these sequences. Since not every topology was examined, it was not possible to assign mutations to connecting branches on the basis of "cost," as for the other topologies reconstructed by the method of minimum mutations. Instead, the minimum mutations were apportioned to the various branches in accord with the following rules (satisfied in that order): (1) that the number of mutations be minimal; (2) that the number of mutations assigned to connecting branches be minimized; and (3) that within the limitations of (1) and (2), fractional mutations be apportioned by giving equal weight to every possible way of assigning the mutations at each nucleotide position.

relevant immunoglobulins (primarily H-chain molecular weight, polymeric form, and overall carbohydrate content) and is subject to radical revision as more crucial data emerge.

Thus, for example, H-chain molecular weight, as determined by physical measurements, is prone to error (Dorrington and Tanford, 1970) and is a relatively uninformative characteristic of $C_H$ genes. The reported sizes of H chains are not entirely consistent with a simple classification: $\gamma$ chains and human and mouse $\alpha$ chains fall into one relatively small class size (about 450 amino acids), while $\mu$ chains, rabbit $\alpha$, and elasmobranch H chains seem to fall into another (about 600 amino acids). The size differences might reflect multiple genetic events which are not the same in all cases. For example, in those portions where their sequences can be compared, $C_\mu$ and $C_\gamma$ have at least two large gaps relative to each other: The last 19 residues of $C_\mu$ are deleted in $C_\gamma$ (see Table 4-1), and another gap is clearly required to account for the reported fact the $C_\mu$ is about 150 amino acids longer than $C_\gamma$.

The polymeric form is another characteristic from which phylogenetic conclusions must be drawn with caution. A single class of immunoglobulin in elasmobranchs, similar in several gross characteristics to mammalian IgM, exists in both a monomeric form (like mammalian IgG) and a (probably) pentameric form (like mammalian IgM); while *Xenopus* (a frog) has a hexameric immunoglobulin and bony fish a tetrameric one (see Table 2-6).

Three major classes of immunoglobulin, differing in their H chains but not in their L chains, have been described in a wide variety of mammals. Since there is every reason to believe that all H chains draw their $V_H$ regions from a common pool of genes (see Chapter 9), it is presumed that the differences between these classes reflect differences in their $C_H$ regions. The overall characteristics of these classes correspond in several respects in different mammals: for example, IgA's have a high carbohydrate content and are characteristically found abundantly as dimers associated with a "secretory piece" in secretions such as colostrum; IgM's have a large H chain (about 600 amino acids), a high carbohydrate content, and are characteristically found as pentamers in the serum. IgG, the predominant serum class in all mammals and for which a great deal of sequence information is available, has a low carbohydrate content and an H chain of about 440 residues. It is accordingly presumed that the $C_\mu$, $C_\alpha$, and $C_\gamma$ genes diverged before the mammals diverged from one another, giving rise to three duplicated genes (or families of related genes) in the common ancestor of all the mammals. This conclusion is supported by the demonstration of immunological cross-reactions between corresponding classes of immunoglobulin in different mammals. A similar conclusion applies to IgE, since both rabbits and men have an antigenically related class with the special characteristics of "reaginic" antibody. There is no known counterpart to human IgD in other species.

In amphibians both a monomeric and a polymeric (probably hexameric) immunoglobulin class have been found in the serum. The former shares with mammalian IgG a relatively small H chain, its monomeric form, and its relatively low carbohydrate content. Its H chain differs by both antigenic analysis and peptide mapping from the H chain in the ploymeric class; the latter resembles mammalian $\mu$ chains in apparent molecular weight and carbohydrate content. It is thus not unreasonable to suppose that the monomeric and polymeric classes of amphibian immunoglobulin share common ancestors with mammalian $C_\gamma$ and $C_\mu$ respectively. This conclusion is supported to some degree by the existence of antigenically distinct H chains in monomeric and pentameric classes of bird immunoglobulin. However, the size of the H chain in the monomeric class seems to be more similar to mammalian $\mu$ chains than to mammalian $\gamma$ chains, casting some doubt on the conclusion (see Table 2-6).

If, as argued above, $C_\mu$ and $C_\gamma$ diverged before the mammals and amphibians did, this divergence probably occurred after the phylogenetic divergence of elasmobranchs from the higher vertebrates; for in the shark-like fishes the monomeric and polymeric immunoglobulins have H chains which are indistinguishable from one another by peptide mapping and antigenic analysis. Therefore the $C_H$ regions of the elasmobranchs might derive from the common ancestor of both mammalian $C_\mu$ and $C_\gamma$. The overall similarities of elasmobranch $C_H$ to mammalian $C_\mu$ would in that case reflect the characteristics of a common ancestor of all three, from which mammalian $C_\gamma$ has diverged more extensively (in these respects) than the other two. Since there are conflicting reports on whether a monomeric class of immunoglobulin is present in the bony fish (see Table 2-6), we are unable to decide whether the $C_\mu$-$C_\gamma$ divergence inferred above occurred before or after the phylogenetic divergence of these animals from the higher vertebrates.

It has been argued on the basis of their similar C-terminal sequences (see Table 2-4B) that mammalian $C_\mu$ and $C_\alpha$ are more closely related to each other than to $C_\gamma$. It is equally possible, however, that $C_\mu$ and $C_\alpha$ merely reflect the structure of the common $C_H$ ancestor in this region, and that $C_\gamma$, after diverging from one or the other of them, lost that region by deletion. We thus cannot say in what order $C_\mu$, $C_\alpha$, and $C_\gamma$ diverged. To my knowledge, no critical search for IgA-like immunoglobulin has been attempted in nonmammalian species and therefore this order of divergence cannot be inferred from phylogenetic considerations. The relation of $C_\epsilon$ and $C_\delta$ to other C regions cannot yet be determined.

It should be very apparent that, except for the protein genealogies presented at the beginning of this chapter, any scheme for C-region evolution rests on shaky grounds and is subject to extensive revision by more crucial data. Nevertheless, Fig. 4–3 presents a tentative descent for the immunoglobulin C regions. Firmly established portions of the descent, which are based on the mutation patterns in sequences, are represented in solid lines; the other less firmly established relationships are represented in dashed lines.

The divergence between H and L chains, and the divergence between the two forms of L chain—κ and λ—represent the duplication of entire gene families, each of which consists of a group of linked, multiple V genes and C genes. The evidence strongly suggests that any of the V genes can be found in association (by some form of somatic joining) with any of the C genes in the same family (and vice versa), as will be explained in Chapter

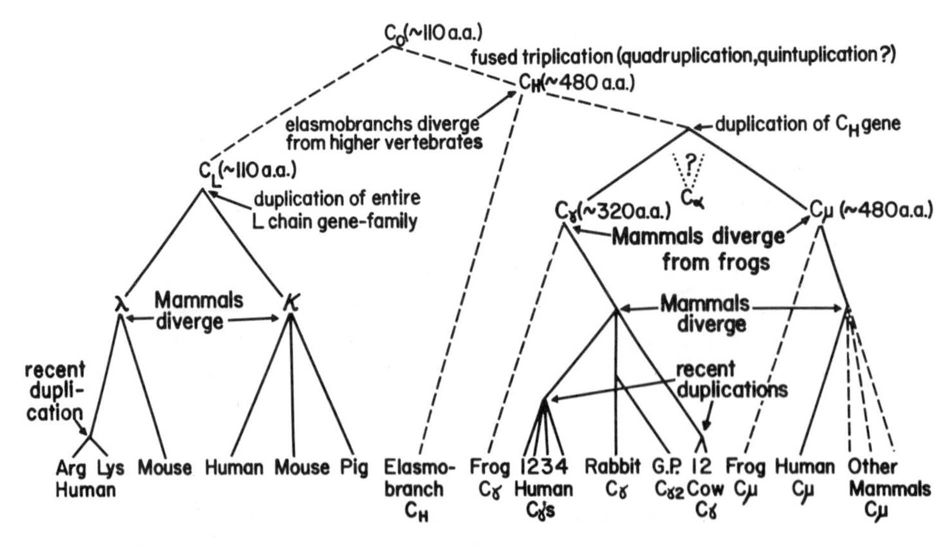


**Figure 4–3.** Composite genealogy, summarizing the descent of C regions. Solid lines indicate genealogies based on Figs. 4–1 and 4–2; dashed lines indicate less certain portions of the descent based on indirect evidence; *a.a.* indicates amino acids.

9. In both their V and C regions, the $\kappa$ and $\lambda$ chains are much more similar to each other than to the corresponding region of H chains: the divergence of these two families therefore is presumed to have succeeded the divergence of their common ancestor from the H-chain family. These families appear to be very important functional units at the level of the genome.

Within each of the families V and C genes have undergone separate duplications. V-gene duplications, which will be discussed in Chapter 6, are assumed to be responsible for the diverse binding specificities of antibodies with diverse antigens. The variants produced by C-region duplication have largely unknown function. The divergences of the duplicated $C_H$ genes contrast with those that still survive among the $C_\kappa$ and $C_\lambda$ genes in that they are older and have led to very distinct classes which are present in all mammals, and probably some nonmammals as well. It is reasonable to suppose that these perform certain highly adaptive specialized functions; for example, the $C_\alpha$ gene might have evolved so as to be able to cross several body barriers and provide immunity in the secretions. It is also reasonable to suppose that one of the surviving $C_H$ regions in mammals continues to perform many of the same functions originally performed by the $C_H$ ancestor. The $C_\mu$ region is a logical candidate, since IgM shares several gross features (high carbohydrate, ploymerizability, large H chain) in common with the immunoglobulins in primitive vertebrates which appear to have but a single class of $C_H$ region. One such function might be to act as the receptor site on antigen-reactive cells (see Chapter 8).

These ideas are highly speculative, but whether or not they are correct, it will never be found (despite numerous statements to this effect in the literature) that IgM is "earlier on the evolutionary scale" than IgG or other immunoglobulins, for the very good reason that all immunoglobulins come from contemporary species and are therefore precisely as old as one another when measured from their point of common origin. Indeed, unless one espouses a most unfashionable view of evolution, one cannot talk meaningfully of the "appearance" of a gene at all, but only of its divergence from other genes with which it shares a common ancestor.

# Chapter 5

## Theories of Antibody Diversity

Three basic mechanisms have been proposed for the origin of antibody diversity: recombination, mutation, and variable translation of ambiguous codons in V-region structural genes. Mutational theories are further subdivided according to whether the mutation is thought to occur *somatically* in one inherited structural gene, or *presomatically* in multiple inherited structural genes. Thus four major classes of theories are now current: recombination, somatic mutational, germline (presomatic) mutational, and translational. In this chapter, I shall discuss in detail the proposals in each of these categories and eliminate the translational theory from consideration. In the next two chapters, the evidence for and against the other three will be discussed at length.

### The Simple Translational Theory

It is a theoretical possibility that a given mRNA codon can specify several different tRNA's, charged with different amino acids. It is also possible that a single tRNA can be charged with several different amino acids, depending upon which of several hypothetical amino acid activating enzymes acts upon it. Both mechanisms lead to the possibility of different amino acid sequences appearing among the products of a single structural gene. Potter, Appella, and Geisser (1965) and Mach, Koblet, and Gros (1967) have suggested that many different variants could be produced from a single antibody structural gene containing such variably translated codons if different sets of tRNA's or activating enzymes were active in different antibody-producing cells. If the further hypothesis is proposed that the ambiguous codons are confined to the V region, the constancy of the C region is also explained.

Tissue-specific differences in activating enzymes have been found and proposed as a mechanism for cellular differentiation (Strehler, Hendley, and Hirsch, 1967). Mushinski and Potter (1969) and Yang and Novelli (1968) have reported differences in the tRNA's in different mouse plasma-cell tumors producing different $\kappa$ chains. There is no evidence, however, that these observed differences have anything to do with variability, and on both experimental and a priori grounds such a relationship seems unlikely.

First, as pointed out by Gray, Dreyer, and Hood (1967), such a mechanism cannot explain the deletions and insertions which occur frequently within each of the V-region families, and thus it certainly does not account for *all* antibody diversity (see Tables 2-2A, 2-3A, 2-4A). Secondly, a translational mechanism predicts that within the family of proteins coded by one variably translated gene, strictly parallel variation will occur at all positions coded by the same ambiguous codon. For example, if the same ambiguous codon codes for variable positions *i* and *j*, one might expect, say, leucine to occupy both these positions in some proteins in the family, isoleucine in others, and perhaps a third amino acid in others; but one would not expect to find any proteins in the family in which *different* amino acids occupy positions *i* and *j*. The fact is that there are no extensive families of immunoglobulins in which this type of parallel variation occurs. I have searched for it exhaustively among the very closely related human $V_\kappa$ sequences belonging to the $V_{\kappa I}$ subgroup, among which no deletions or insertions have been demonstrated. There are 54 positions which show variation in these proteins, and among them there is not a single pair which vary in parallel; that is, every possible pair of variable positions is different in at least one protein. According to the translational theory, each of these must therefore be coded by a different ambiguous codon; added to the 19 necessary (at a minimum) to code for the invariant positions, they would far exceed the 64 possible codons.

The translational theory can be rescued from these considerations by introducing new complications. If it is supposed that there are *several* genes for $V_{\kappa I}$, each of which can give rise to a family of translational variants, then the variation at many of the 54 variable positions can be attributed to conventional germline mutation during the divergence of these several genes, and the total number of codons required reduced to a possible number. But while this version of the translational theory cannot be excluded simply because it requires too many codons, it must be borne in mind that any translational theory demands a most unlikely concatenation of changes in the genetic code in order that the variability so introduced not destroy the other machinery of the cell. Aristotle advises right-thinking scientists to prefer the probable impossible to the improbable possible. Here our choice is easier. In comparison to the three other "probable possibilities" to be discussed in this chapter, the translational theory is an "improbable possible" which we may justifiably ignore.

A different translational theory has been proposed by Haurowitz (1967). In contrast to the simple theory, the choice between several possible translations of a given codon is made not by tRNA's or activating enzymes, but rather by the antigen, which interacts with the nascent antibody polypeptide chain and favors incorporation of one amino acid over others. While such a mechanism could conceivably change the frequencies of incorporation of different amino acids at different positions, it is difficult to believe that an antigen template could account for a single amino acid sequence, such as is found in myeloma proteins.

## The Somatic Mutation Theory

> Then Possibility No. 3 is knocked on the head. There remain Possibilities No. 1 and 2. Following the methods inculcated at the University of which I have the honour to be a member, we will now examine severally the various suggestions afforded by Possibility 2. This Possibility may again be subdivided into two or more Hypotheses. On Hypothesis 1 (strongly advocated by my distinguished colleague Professor Snupshed), . . .
>
> —Dorothy L. Sayers (1923): *Whose Body?*

Between 1957 and 1959, Burnet (1959) developed a hypothesis—the clonal selection theory—which revolutionized theoretical immunology. In one stroke it provided natural, economical explanations of the signal features of the antibody response: the specificity of antibody production, the secondary response, and tolerance to self-antigens.

The clonal selection theory was modeled after bacterial genetics. Like antibody diversity, bacterial variation, which results from random mutation, is preadaptive. Burnet proposed that during cell proliferation in the development of the immune apparatus, antibody structural genes undergo similar random mutation, so that different clones of cells result, each clone committed to the production of a particular variant of antibody; somatic mutation accordingly was posited as the origin of antibody diversity. Those clones committed to antibody reactive to endogenous antigens would be suppressed, and thus a purge of inappropriate clones was proposed as the explanation of tolerance. An exogenous antigen, however, would (according to the theory) have an opposite effect on cells which happen to harbor a variant of antibody reactive with it: these cells would be stimulated to proliferate and produce antibody, thus accounting for the synthesis of specific antibody on immunization. The mechanism of this selective stimulation (and of selective suppression by endogenous antigens), though left unclear in detail, was thought to depend upon a primary combination of the antigen with small amounts of the antibody produced constitutively by the cell. In this way antigenic stimulation (or suppression) would naturally be selective for cells committed to antibody of corresponding specificity. Some of the cells stimulated to proliferate by a primary contact with exogenous antigen would not (according to the theory) immediately produce antibody, but rather wait, in great numbers, for a secondary contact with antigen: the accelerated and enhanced antibody response generally observed to a second immunization with antigen thereby is explained. These are called the memory cells, since they are hypothesized as the means by which an animal "remembers" previous contact with a given antigen; they are presumably the basis of the commonly observed immunity to second attacks of many infectious diseases.

Until about 1965, Burnet's clonal selection theory and the particular mechanism of variation he proposed (somatic mutation) were often considered inseparable. By now it is clear, however, that any of the current hypotheses of antibody variability can be incorporated naturally into the clonal view of the antibody response. It is, I think, in the clonal view of immunity—not in the somatic mutation mechanism—that the elegance and economy of Burnet's theory reside. Therefore, in the remainder of this book, the clonal selection theory is intended to mean only the theory of heritable cellular commitment, and will not (and need not) assume any particular mechanism of generating variability. The clonal selection theory, in this restricted sense at least, is by now almost a central dogma and received truth of immunology, for without its insight very little of the phenomenon of immunity is comprehensible. This important but relatively uncontroversial aspect of immunology will be dealt with separately in Chapter 8, and we shall return now to a detailed examination of somatic mutation as the possible origin of antibody diversity.

It is clear that a viable somatic mutation theory must in some way provide for *hypervariation* of antibodies; for somatic mutation, if it occurs at all in other genes, does not lead to a comparable heterogeneity of other proteins. Somatic mutation theories therefore incorporate either some mechanism of *hypermutation* or some *somatic selection process* by which somatic mutations are expressed more frequently in antibodies than in other proteins. Secondly, somatic mutation theories must account for the restriction of expressed mutations to the V regions only of immunoglobulin polypeptide chains.

Brenner and Milstein (1966) proposed a somatic mutation theory in which hypervariability is caused by hypermutation. An enzyme repeatedly degrades one strand of the antibody gene, and during its repair (by replication off the other strand) errors are introduced. While the introduction of errors during one repair would not necessarily be high, the high frequency of repair necessitated by the degrading enzyme would lead to a high frequency of mutations. Both point mutations and deletions or insertions could be accounted for in this manner. Since no hypervariation is observed in other proteins nor within antibody C regions, these authors suggested that the degrading enzyme recognizes V-region DNA by some specific recognition site.

Other somatic mutation theories hypothesize that hypervariation results from *selection* for mutations, the mutations themselves occurring at a more or less normal rate. The selection agent is usually thought to be antigen of some sort, which asserts its influence by combining with the corresponding antibody combining sites. Since the characteristics of the antibody combining site are thought to depend on the structure of the V regions of which it is composed, V-region mutations will in general be selected for; although mutations in C regions would presumably occur as often as those in V regions, only the latter will finally be expressed with detectable frequency.

Several such selective versions of the somatic mutation theory have been proposed. Cohn (1968, 1970) proposed that exogenous antigens select for antibody mutants by stimulating those clones which are committed to antibodies reactive with them. Since the chance that an unmutated antibody will react with a given exogenous antigen is low, most stimulation will occur in cells producing mutant antibodies, and hence a constant low level of antigenic stimulation might produce a highly heterogeneous population of cells in which mutants predominate. If, then, it is supposed that any given antigen (whether or not it participated in the original selection) cross-reacts with some proportion of the cells, this process will generate sufficient antibody diversity to explain immunocompetence.

Jerne (1971) proposes, on the other hand, that "nature exploits the fact that the most powerful selection pressure favoring mutants is the suppression of nonmutants." Inherited antibodies, he supposes, are directed against self-antigens, particularly the histocompatibility antigens. They have been selected in evolution for this feature, according to Jerne, perhaps in order to facilitate constant surveillance against neoplastic mutants (which often express new surface antigens). The same need necessitates a great heterogeneity of histocompatibility antigens, and since no animal "knows" in advance which of these antigens he will inherit, he inherits antibodies against all of them. But since, in order to survive, the animal must be tolerant to those histocompatibility antigens which he happens to inherit, the same mechanism of clonal purge, which is in any case the most plausible explanation for immunological tolerance to other endogenous antigens (whether natural or experimentally introduced), would be expected to suppress those clones of cells producing antibody to endogenous histocompatibility antigens. Mutants of these antibodies, however, might be expected to react poorly with these antigens and escape suppression. In this way, without any epiphenomenon being introduced into his theory, Jerne might account for a sufficient variety of mutant potential antibody-producing cells to provide for general immunocompetence against exogenous antigens.

Clearly a great many other variations of this selective somatic mutation theory are possible, and it seems to me that they are at least as plausible as the alternative theories. Several theoretical objections might be raised to them (as well as to any of the other theories), but since the different theories predict distinct outcomes of several experimental approaches, it does not seem profitable to dwell on such unproven considerations.

**The Germline Theory: The Problem of Selection in Antibodies**

> Ah-hem! Trusting, gentlemen, that you have followed me thus far, we will pass on to the consideration of Hypothesis No. 2, to which I personally incline. . . .
>
> –Dorothy L. Sayers (1923): *Whose Body?*

The theory that each variant of antibody is coded by a separate inherited structural gene would hardly be a novel view if antibody diversity were not apparently preadaptive and very large. Variants in other families of proteins–for example, the $\alpha$, $\beta$, $\gamma$, and $\delta$ chains of human hemoglobin–are all no doubt coded by separate germline genes. But each of these variants presumably performs in every animal some particular function for which it has been individually selected in the course of evolution. What selective force can be invoked to explain the maintenance in the germline of the ability to respond specifically to arbitrarily chosen, even artificial, antigens which could have played no significant role in the evolutionary past of the immunocompetent species?

In a careful study of the pattern of variation in L chains, Wu and Kabat (1970) have provided some insight into the nature of the selective forces acting on these genes. The amino acid positions can be divided into three broad categories: invariant or nearly invariant positions, normally variable positions, and hypervariable positions. The unevenness of the distribution precludes the hypothesis that mutations accumulate (and are accepted) randomly, and demonstrates that some selective force is operating on antibody genes.

There are two possible interpretations of the hypervariable positions. On the one hand, they might be positions at which little selection is operating, and therefore they are free to accumulate mutations at random. On the other hand, as Cohn and his co-workers (Cohn, 1970; Weigert et al., 1970) have urged, they might be positions at which many substitutions confer a positive selective advantage, and therefore they are even more variable than expected in the absence of selection. At least two of the hypervariable regions found by Wu and Kabat are close to the V-region intrachain disulfide bond and therefore to each other. This fact, and the preferential labeling of nearby residues by affinity labels, have led to the conclusion that the hypervariable regions constitute the antigen binding sites of antibodies. It is reasonable, therefore, that they be under a special sort of selective pressure and that this pressure lead to a diversity of sequences in these segments.

The evidence for selection in antibody genes poses some difficulty for the germline theory. This difficulty is not so great, however, as is often supposed. It is frequently argued that the germline theory requires the maintenance of an almost unlimited number of genes specific for rarely encountered, even artificial, antigens, even though they are seldom if ever called upon to act. How can natural selection maintain such evolutionarily silent genes in a functional state? The key assumption of this argument is that an antibody gene is highly specific for some corresponding antigen. This assumption is not justified by the facts of serological specificity. Briefly, with Talmage (1959) I believe that it is quite plausible to consider that the specificity of an immune response resides in the appropriate combination of individually relatively nonspecific elements. These combinations occur at three different levels: an anitgen is a combination of antigenic determinants, "antibody" is a combination of molecular species (all of which bind to some determinant on the antigen), and an individual molecule of antibody is a combination of two chains, each of which might contribute to the specificity of the whole molecule.

It has become increasingly clear that antibody elicited by most antigens is highly heterogeneous. Partly this heterogeneity arises from a heterogeneity of antigenic determinants on single complex antigenic molecules: but even single synthetic haptens elicit a heterogeneous mixture of antibody molecules. The heterogeneity of antibody raises the possibility that individual antibody molecules are *not* highly specific, the specificity of a given antibody preparation being rather a property of a whole complex mixture of molecular species whose sole common feature is that they all happen to bind to some determinant on the eliciting antigen. Analogously, the number of distinguishable antigenic determinants might be limited, the uniqueness of each antigen being attributable to the unique combination of antigenic determinants it represents.

Lastly, even the specificity of individual molecules of antibody might result from the combined characteristics of their H and L chains, either of which, in different combinations, might determine quite different specificities. The evidence (Dorrington, Zarlengo, and Tanford, 1967; Grey and Mannik, 1965; Mannik, 1967) that the actually expressed pairs of H and L chains have a higher affinity for one another than randomly chosen combinations of chains implies that a given chain cannot be expressed in combination with *any* partner. But even if this restriction is quite severe (as few as 1 percent of the possible combinations being functional, for example), a large number of different specificities might result from a limited number of original antibody sequences.

It must be remembered that an immunocompetent animal has been negatively immunized with his own antigens and does not produce antibodies which react with them. Hence an animal must be able to produce antibodies with sufficient specificity that, when those which react with his own antigens are subtracted, those that remain can account for general immunocompetence to foreign antigens. Even taking into account the demand of tolerance for specificity, however, the "combinatorial" hypothesis remains a reasonable explanation of how highly specific responses to a limitless number of antigens might be mounted with a limited repertoire of different antibodies.

If, then, individual antibodies (and their compenent polypeptide chains) are relatively nonspecific, there is no reason to believe that they are evolutionarily silent. On the contrary, they would probably be expressed frequently, because each one could be elicited by a wide range of antigenic determinants, possibly including some artificial haptens that have never appeared in nature.

In eliminating the necessity for assuming an unlimited number of silent genes, however, the combinatorial view of specificity introduces another, though less serious, evolutionary difficulty for the germline hypothesis. The relative lack of specificity of the individual antibody genes implies that they are in large measure redundant. Indeed, the heterogeneity of the antibody produced against even single antigenic determinants makes the redundancy of antibody diversity an almost unavoidable fact.

This redundancy would be a natural outcome of somatic mechanisms of generating diversity, for in the course of generating some variants reactive to all antigens, they must be supposed to generate many variants reactive to most antigens. And if, as in some somatic models, selection in the individual amplifies these variants, the selective force operating on one variant is independent of that operating on another; consequently all the redundant variants meeting the requirements of some given somatic selective force will survive in the individual.

In the germline theory, on the other hand, redundancy must be maintained by a selective force which does not operate independently on each individual variant, but rather can only select on the basis of the overall immunological performance of a given genomic complement of genes. Even if we admit that all the putative antibody genes are expressed frequently, and thus are subject to the constant surveillance of natural selection, could this selection be powerful enough to maintain in a functional state a large set of more or less redundant genes, a loss of any one of which could hardly have a large effect on the overall effectiveness of the genome?

If the combinatorial view outlined above is correct, the actual redundancy of antibody genes is much less than appears on first consideration. For although (in that view) any one gene contributes only a tiny fraction of the overall response to one antigen, it contributes similar fractions of the responses to countless other antigens as well, and the sum of these contributions may amount to near indispensability in the race for survival. The evidence, however, is that natural selection is somewhat less solicitous of individual genes: V genes are constantly being deleted in various species and replaced by duplications of other V genes (Chapter 7).

As will be discussed in Appendix C, there are two well-studied systems of multiple genes in higher organisms—the ribosomal RNA genes and the 5s RNA genes—whose redundancy (unlike that of antibodies) is not in question. Despite their redundancy, they are highly conserved in evolution. Unfortunately, these two systems are less than perfect precedents for the superficially analogous case of multiple germline antibody genes. Neither of the two most widely held theories to account for the conservation of ribosomal and 5s RNA genes adequately explains the pattern of variation evinced by the immunoglobulins. The first, and in my opinion most likely, theory is rapid gene turnover resulting from unequal crossover; it has the defect of predicting that the heterogeneity of all positions in the repeated genetic unit be approximately the same, which, as we have seen, does not obtain among the immunoglobulins. This evidence by no means precludes that immunoglobulin genes turn over by unequal crossover. Indeed, as will be discussed in Chapter 7, this turnover is a very plausible explanation for the evolution of species-specific differences in V regions. What the uneven distribution of heterogeneity does do is greatly weaken gene turnover as an explanation for the maintenance of redundant V genes in a functional state by a reasonable selective force that does not greatly care about the loss of any one of them.

The second theory is the master-slave hypothesis, which postulates that all the redundant copies in a multigene family are made identical to a single master copy at every meiosis; this postulate, if applied to antibodies, would throw the baby of diversity out with the bath-water of deleterious mutations.

Gally and Edelman (1970; Edelman and Gally, 1970) have put forward a possible explanation for the maintenance of redundant antibody genes in a functional state. Although proposed in the context of a somatic recombination theory of diversity, their "democratic" gene conversion model serves equally well in the context of the germline hypothesis. They suggest that any one of a set of multiple genes can convert a portion of a nearby gene to its own sequence. Deleterious mutations can be thereby rapidly corrected without discarding the essential diversity, and favorable mutations can be spread to neighboring genes in the set. Partial gene conversion is formally equivalent to recombinations involving short stretches of sequence. As will be discussed in the next chapter, this recombination would be expected to lead to a high level of dislocations, which would ordinarily be interpreted as parallel and back-mutations. Dayhoff and her colleagues (Dayhoff, 1969), however, failed to find a higher incidence of apparent parallel and back-mutations in antibodies than in other families of proteins, casting considerable doubt on the democratic gene conversion theory.

Although its worst problems are obviated by the combinatorial theory of antibody specificity, therefore, the germline theory has some difficulty accounting for the maintenance of functional, apparently redundant diversity by ordinary natural selection. Nevertheless, this consideration can hardly be a decisive one. The conclusion that a germline theory is evolutionarily impossible is removed by too many steps of inference and unsure (though plausible) assumption from observable facts to settle the issue, and we must rely on much more direct evidence of the sort to be discussed in the next two chapters to decide the question finally.

The demonstration (Hilschmann and Craig, 1965) that the antibody variation which might account for specificity is confined to the V regions of immunoglobulin chains has introduced complications for most theories of antibody diversity. Somatic mutation theories have to provide some way to restrict the expressed somatic mutations to V regions. Those based on hypermutation have to suppose that the hypermutating apparatus recognizes only V regions; those based on somatic selection are least affected, since the selective agent may be assumed to act on subtle properties of the combining sites, which can plausibly be said to be determined solely by V regions.

But for the germline theory, the contrast between V and C regions was a paradox. It supposed multiple ancestral duplications of antibody genes, but the observed constancy of C regions was difficult to reconcile with their determination by ancestrally duplicated cistrons. It could only be rescued by the heretical expedient of separating in the germline the genes determining V and C portions of the immunoglobulin chains—V genes being present as multiple variant copies, but each C gene being unique. The separation of V and C genes in the germline is thus an essential feature of the germline hypothesis, but not of the theory of somatic mutation. It has now been proved by a different sort of reasoning that V and C genes are not associated with one another on a one-to-one basis in the germline (Chapter 9), and the proponents of somatic mutation must introduce into their theory a gratuitous mechanism for joining V and C genes or their products.

Dreyer and Bennett (1965) proposed the current germline theory and it has not been significantly modified since (Dreyer and Gray, 1968; Dreyer, Gray, and Hood, 1967). They pointed out that if the joining of V and C regions is accomplished at the DNA level, only one of the multiple V genes can be joined to a given C gene in any one cell, and in this way a heritable clonal commitment to the expression of a single one of the V genes is effected. The germline theory of diversity, then, can be incorporated (as any reasonable theory must) into the clonal view of adaptive immunity as naturally as can the hypothesis of somatic mutation, which formed part of Burnet's original clonal selection theory.

In all major respects, the conservative view that all diverse antibodies are the directly expressed products of correspondingly diverse inherited genes is at least as plausible as other ways of accounting for antibody diversity. The argument that it cannot account for selection in antibodies rests on too many unsure assumptions to be compelling; and the only really novel epiphenomenon invoked by the theory—the somatic joining of V and C genes separated in the germline—has been proved to occur in any case: one cannot prosecute for a heresy which God Himself embraces.

**The Somatic Recombination Theory**

The virtue of a somatic mechanism of generating antibody variation is that the potential for diversity, limited only by the number and rate of proliferation of lymphoid cells, is incomparably larger than that which could be maintained in the germline. Once evolved by selection, such a mechanism would naturally produce the apparent redundancy which poses some difficulty for the germline theory. Furthermore, the so-called "hypervariation" mechanism may be nothing more than the selective stimulation and suppression of cells by antigens, which must anyway be postulated to explain adaptive immunity.

If somatic variation is by mutation, however, nowhere near the full potential for diversity can be usefully exploited, for many cells would be wasted because of nonfunctional variants. If, on the other hand, somatic variation results from recombination among a small set of germline genes, the potential for diversity inherent in the lymphoid cell population might be much more fully exploited, especially if the genes were optimized by natural selection for this result. Somatic recombination merits careful consideration as a possible mechanism of producing antibody diversity.

Since the recombination theory was first proposed by Smithies (1963), it has had to be modified extensively to conform to the data. When the first sequence data revealed that only V regions harbored extensive variation, Smithies (1967a and b) proposed that recombination occurs between a complete VC gene and a half-gene for the V region. As more amino acid sequence data accumulated, however, it became clear that any minimal hypothesis involving recombination among a highly restricted number of V genes (three or four, for example) is untenable, for the sequences cannot be accounted for as relatively simple recombinants of a few linkage groups. More recently, Gally and Edelman (1970), in order to reconcile the theory with some of the considerations in the next chapter, proposed that recombination occurs among a few (about ten) V genes within each of several subgroups in each gene family, but not between V genes in different subgroups.

In order to test the current version of the somatic mutation theory adequately, therefore, it will be necessary to devise an assay which does not depend on the participation of only a few germline genes and which can detect recombination within subgroups of closely related V regions. Such an assay will be outlined in the next chapter.

It should be noted that *any* sequence can be explained as a multiple recombinant of four genes (for example, poly A, poly T, poly C, and poly G). Recombination becomes a plausible explanation only if it is relatively simple, requiring only a few crossover events per chain.

# Chapter 6

## The Evolution of V Regions: The Case for the Germline Theory

A detailed analysis of the pattern of variation in myeloma proteins constitutes what I consider to be the most direct evidence on antibody diversity. The debate that has arisen over the meaning of this evidence has been unnecessarily complicated and has given rise to several distortions and misconceptions. These difficulties can be largely avoided once it is recognized that almost every important inference drawn from this information has rested (at least implicitly) on genealogical reasoning. The data can be handled more exhaustively and at the same time more economically by starting afresh from genealogical principles, which will be used here to develop a systematic analysis of the descent of the immunoglobulin V regions. This descent will in turn be compared to the contrasting expectations of the major theories of antibody diversity.

In effect this project has been a search for more and more sophisticated techniques to detect and display the anomalies to be expected if either of the two somatic theories is correct; for if the germline theory is right, the pattern of variation in V regions should resemble in every particular that observed in other families of evolutionarily related proteins. I have not found such anomalies, and my data join those of others in putting restrictions on the somatic theories which reduce considerably their appeal as explanations of antibody diversity. To these considerations must be added the special case of identical V-region sequences from different individuals: these virtually prove that the basic tenet of the germline theory—that an immunoglobulin is a faithful translation of the germline gene from which it derives—holds in at least some cases. However, there is a countervailing body of evidence which is awkward for the germline theory; it will be discussed at length in the next chapter.

### Expectations of Theories of Antibody Diversity

#### *The Somatic Mutation Theory*

The expectations of this theory are diagrammed in Fig. 6-1, which shows the genealogical relationships to be expected among six myeloma proteins which all derive by somatic mutation from the same germline gene. The

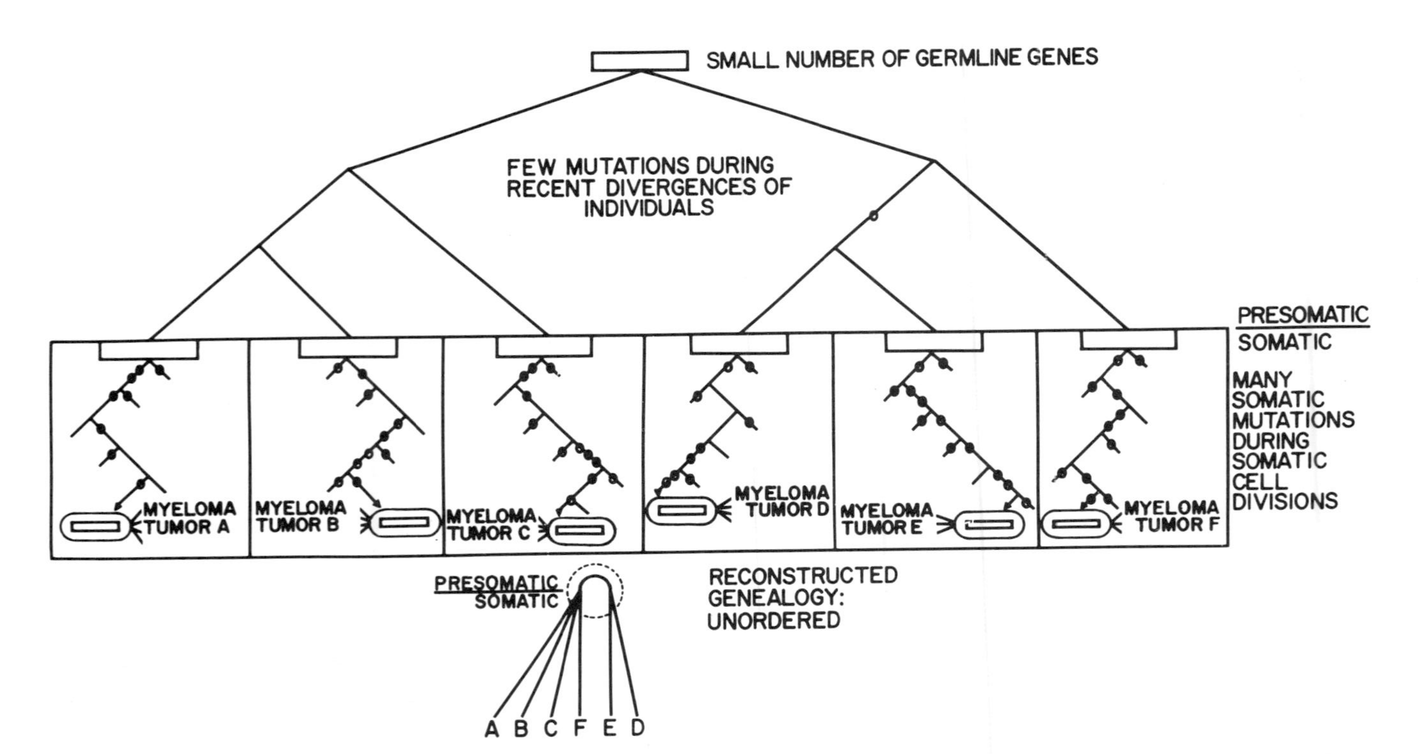


**Figure 6-1.** Expectations of the somatic mutation theory. The six myeloma proteins shown are somatic variants of a single germline gene. Since almost all mutations occur somatically, few can be shared by proteins from different individuals, and the reconstructed genealogy is therefore unordered.

mutations in these six proteins have by hypothesis occurred somatically, and hence independently in each individual. Accordingly, they should be related by an unordered genealogy, all of them emanating (within the uncertainties of reconstruction) from a single node.

Conversely, if a set of myeloma proteins is related by an ordered genealogy, it is highly unlikely that they derive by somatic mutation from identical germline genes. The mutations assigned to the connecting branches in such a genealogy are shared by at least two proteins from different individuals, and unless one is willing to suppose that the proteins suffered these mutations independently in parallel, they must have occurred presomatically. This principle is illustrated in Fig. 6-2, which also demonstrates how a minimum number of different germline genes can be enumerated from a given genealogy: this number is equal to the number of separate nodes in the genealogy from which a terminal branch (that is, one terminating in an actual sequence) emanates.

While an ordered genealogy allows one to deduce a minimum number of different germline genes, it says nothing in itself about how these genes are located on the chromosome. Many of the germline genes inferred by genealogical methods might logically be supposed to be allelic to one another. Genealogical evidence of this sort cannot rule out the somatic mutation theory, even if highly ordered V-region genealogies are found. Such genealogies merely increase the number of germline genes which somatic theories must accommodate, and thus reduce their economy of explanation.

*The Germline Theory*

Highly ordered V-region genealogies, implying a relatively large contribution of presomatic mutation, are an expected outcome of the germline theory, as shown in Fig. 6-3, for in contrast to the somatic mutation theory it supposes that all immunoglobulin variation results from mutation in the germline. It should be emphasized, however, that although an ordered genealogy is strong positive evidence for a multiplicity of germline genes, an unordered genealogy cannot be taken as strong evidence against them. It is only mutations which have occurred in those lines of descent corresponding to connecting branches which can be identified as presomatic in the reconstructed genealogy; those occurring in the terminal branches can be either presomatic or somatic. In fact, in matching the actual genealogies with the expectations of the germline theory, we certainly cannot reasonably expect any higher degree of order than that

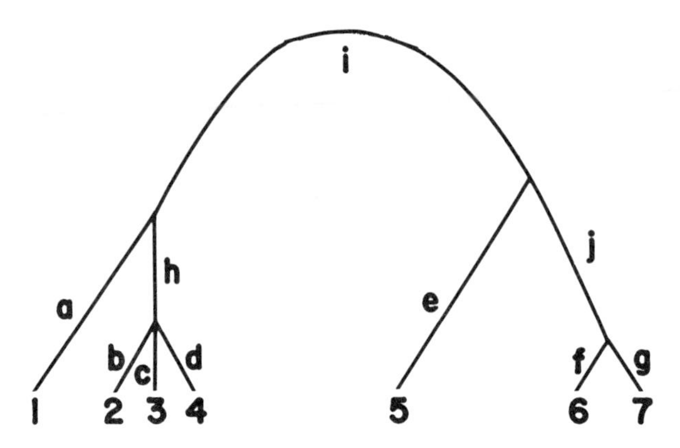


**Figure 6-2.** Enumerating germline genes from a reconstructed genealogy. This hypothetical highly ordered genealogy delineates some mutations which must have occurred presomatically. All the mutations in the terminal branches (a to g) can logically be supposed to have occurred either somatically or presomatically. The mutations in the connecting branches (h to j), however, are all shared by at least two proteins from different individuals; since proteins from different individuals must have diverged from one another presomatically, their shared mutations must also have occurred presomatically (unless we are willing to suppose that they occurred in parallel in those individuals). Thus mutations in connecting branches occurred presomatically. Two nodes separated by a connecting branch with a significant number of mutations must have diverged presomatically from one another and therefore represent two germline genes with different sequences. These germline genes may be either allelic or nonallelic, and they may represent either contemporary germline genes or the common ancestors of contemporary germline genes. The minimum number of contemporary germline genes with different sequences is the number of separate nodes in the descent from which terminal branches emanate; in this case, there are at least four such genes, one each for the 2-3-4, 1-2-3-4, 5-6-7, and 6-7 nodes.

shown in other families of related proteins which everyone assumes to have evolved in the same way that the germline theory supposes for the immunoglobulins. In a genealogy reconstructed for the cytochromes *c*, for example, the proteins from lamprey, dogfish, tuna, and man all appear to have diverged from a single node, even though it is generally acknowledged that these species evolved by an ordered series of divergences (Dayhoff, 1969).

There are two distinct reasons why a reconstructed genealogy might be unordered: there is either a paucity of informative positions, or an abundance of informative positions whose testimony is contradictory. In the former case, we can conclude that few presomatic mutations occurred in those lines of descent corresponding to the connecting branches; if (in accord with the germline theory) we maintain that *all* mutations occur

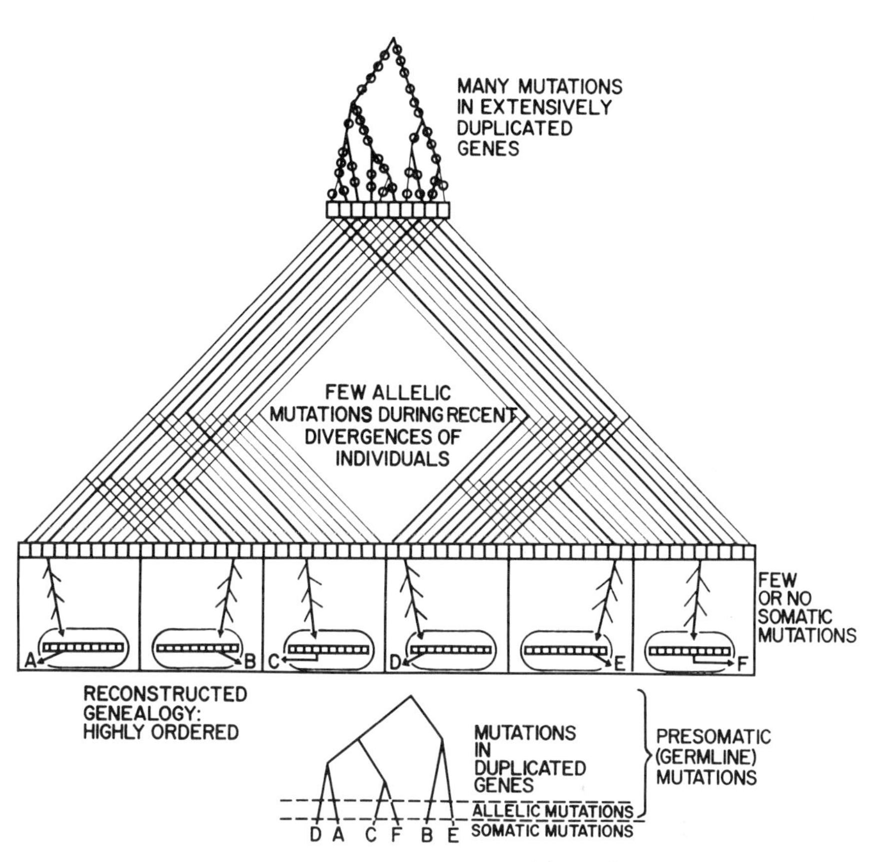


**Figure 6–3.** Expectations of the germline theory. In this simplified example, ten duplicated genes are inherited by all individuals, and each of the myeloma proteins A to F represents a faithful translation of one of them. The reconstructed genealogy, which follows the bold lines in the actual descent, is highly ordered, since many mutations have occurred presomatically and therefore can be shared by proteins from different individuals.

presomatically, their absence from connecting branches is an ad hoc (but by no means unreasonable) assumption. If, on the other hand, an unordered genealogy results from contradictory informative positions, we may conclude nothing about whether or not mutations have occurred in connecting branches; evidently in such cases parallel and back-mutations (and possibly recombination) have obscured the evolutionary history of the proteins. Most cases of unordered genealogies in this book result from contradictory informative positions and thus permit no conclusions; two of them (the $V_{\kappa II}$ and $V_{\kappa III}$ subgroups), however, result from a lack of informative positions and thus constitute at least weak evidence in favor of somatic mutation in these proteins.

In summary, then, an ordered genealogy constitutes positive evidence for a multiplicity of different germline genes and introduces complications into the somatic theory. An unordered genealogy arising from a paucity of informative positions is weak evidence for somatic mutation, since the germline theory must in this case suppose that few if any mutations occurred in those lines of descent corresponding to connecting branches. An unordered genealogy resulting from contradictory informative positions permits no such conclusions.

It is difficult at first to understand the difference between the expectations of the germline and somatic mutation theories. The most frequent counterargument is the following. The only difference between the two theories is that the former attributes to presomatic mutation what the latter attributes to somatic mutation; how, then, can there be any difference between their expected patterns of mutation?

The answer is that if we were to compare several variants from the same individual, we would indeed have no way of ascertaining (by the comparison alone) whether their differences arose somatically or presomatically. But each of the myeloma proteins studied actually arises in a different individual; they have therefore all diverged presomatically from one another. Any mutation which (on the basis of an ordered genealogy) must be assigned to the connecting branches in their descent is shared by at least two proteins and therefore must have occurred presomatically. Two nodes separated by such a connecting branch must consequently represent two distinct germline genes whose differences are caused by presomatic mutation. In this way, a minimum number of distinct germline genes can be inferred, by the rules in Fig. 6-2. These inferred germline genes need not represent contemporary genes, however; if the germline theory is correct, they represent common ancestor genes, from which contemporary genes (and proteins) have derived by gene duplication and germline mutation. Another way of saying the same thing is that there is no way of ascertaining whether mutations in terminal branches are presomatic (as the germline theory would suppose) or somatic.

This reasoning, stated in a more axiomatic form, is exactly the same reasoning that compels us (for example) to suppose that the $\alpha$ and $\beta$ chains of hemoglobin diverged from each other once in the germline and do not arise afresh in each individual by somatic mutation. It is the same reasoning that leads to the universally accepted conclusion that the $V_{\kappa I}$, $V_{\kappa II}$, and $V_{\kappa III}$ subgroups of human $\kappa$ chains are coded by separate germline genes. Both these conclusions rest on the same fundamental premise: that multiple, exactly parallel mutations are highly unlikely to accumulate in independently evolving genes.

Once this abstraction is understood, there is no necessity to argue each case individually. The evidence for presomatic mutation is fully displayed in the form of a genealogy and can be assayed at a glance. Furthermore, the abstraction allows us to draw conclusions on the basis of evidence which might otherwise be considered insufficient. For example, the concept of "subgroups" has hampered the interpretation of V-region evolution. This concept was introduced to describe groups of V-region sequences within one family (such as the $V_{\kappa I}$, $V_{\kappa II}$, and $V_{\kappa III}$ subgroups of the human $V_{\kappa}$ family) which are obviously more similar within groups than between them. Correctly, it was argued that such subgroups must be attributed to separate germline genes. However, it is not necessary to describe a new subgroup in order to define a new germline gene: a single sequence can be sufficient. In the hypothetical example in Fig. 6-2, the sequences 1 and 5 allow two new germline genes to be deduced, by virtue of where they connect to the remainder of the genealogy; it is not necessary to find another 1-like or another 5-like sequence.

*The Somatic Recombination Theory*

The expectations of this theory differ depending on the frequency and multiplicity of recombination events and the number of genes involved. One expectation, however, is common to all the recombination theories, no matter how many genes are involved: recombination should scramble the genealogical relationships in different portions of the immunoglobulin polypeptide chain, as shown in a hypothetical example in Fig. 6-4. A single, simple crossover event often can be reconstructed by genealogical methods. Provided only that the two genes involved are related in different ways to the other proteins under study (and quite independently of the total number of genes), a crossover event will lead to a run of dislocations, very different from the scattered dislocations expected from parallel and back-mutations.

The somatic recombination theory, however, would suppose that *most* proteins are recombinants. Each such recombination event can scramble the genealogical relationships in a different way, and the resulting dislocations should consequently shift from one alternative genealogy to another, leading to a scattered pattern not unlike that expected from parallel and back-mutation. However, because of the resulting genealogical scrambling, no genealogy should stand out as best, and an unordered reconstruction should result. Furthermore, this scrambling should make the good genealogies in the first half poor in the second (and vice versa). A

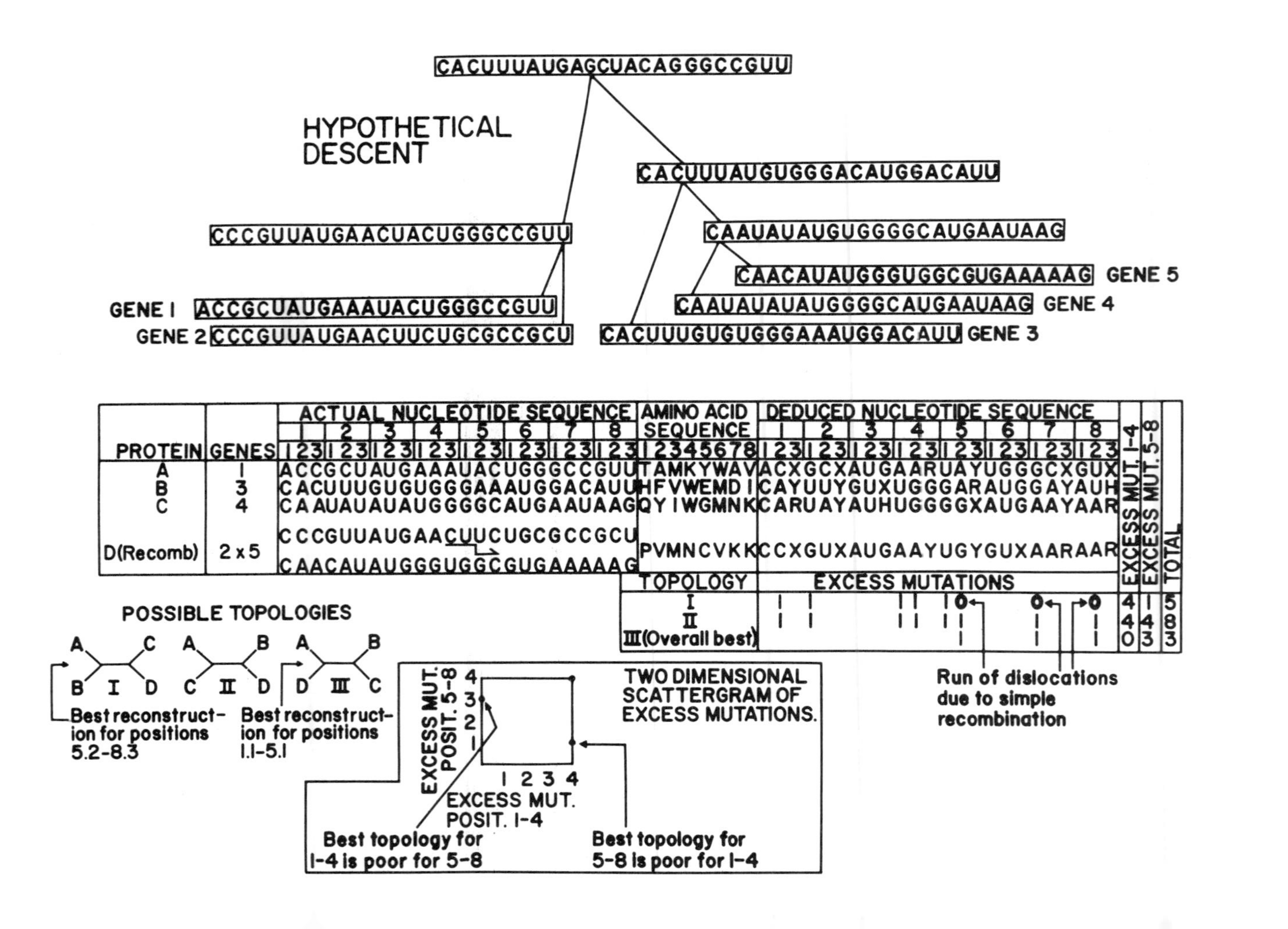

HYPOTHETICAL DESCENT
CACUUUAUGAGCUACAGGGCCGUU
CACUUUAUGUGGGACAUGGACAUU
CCCGUUAUGAACUACUGGGCCGUU
CAAUAUAUGUGGGGCAUGAAUAAG
CAACAUAUGGGUGGCGUGAAAAAG GENE 5
CAAUAUAUAUGGGGCAUGAAUAAG GENE 4
GENE I ACCGCUAUGAAAUACUGGGCCGUU
GENE 2 CCCGUUAUGAACUUCUGCGCCGCU
CACUUUGUGUGGGAAAUGGACAUU GENE 3
PROTEIN
GENES
ACTUAL NUCLEOTIDE SEQUENCE
AMINO ACID SEQUENCE
DEDUCED NUCLEOTIDE SEQUENCE
A I ACCGCUAUGAAAUACUGGGCCGUU TAMKYWAV ACXGCXAUGAARUAYUGGGCXGUX
B 3 CACUUUGUGUGGGAAAUGGACAUU HFVWEMD I CAYUUYGUXUGGGARAUGGAYAUH
C 4 CAAUAUAUAUGGGGCAUGAAUAAG QY IWGMNK CARUAYAUHUGGGGXAUGAAYAAR
D(Recomb) 2 x 5 CCCGUUAUGAACUUCUGCGCCGCU CAACAUAUGGGUGGCGUGAAAAAG PVMNCVKK CCXGUXAUGAAYUGYGUXAARAAR
TOPOLOGY
EXCESS MUTATIONS
EXCESS MUT. I-4
EXCESS MUT. 5-8
TOTAL
I 4 I 5
II 4 4 8
III (Overall best) 0 3 3
Run of dislocations due to simple recombination
POSSIBLE TOPOLOGIES
A C B I D
A B C II D
A B D III C
Best reconstruction for positions 5.2-8.3
Best reconstruction for positions I.I-5.I
TWO DIMENSIONAL SCATTERGRAM OF EXCESS MUTATIONS.
EXCESS MUT. POSIT. 5-8
EXCESS MUT. POSIT. I-4
Best topology for I-4 is poor for 5-8
Best topology for 5-8 is poor for I-4

two-dimensional scattergram of the excess mutations for each genealogy in the first and second halves of the chain should show little correlation, as illustrated in Fig. 6-4. A perfect correlation is not expected even if no recombination occurs (this would imply that each of the intermediate ancestors in the descent acquired equal numbers of mutations in each half), but if the best genealogy (ies) in one half is (are) also the best in the other, multiple recombination events are most unlikely. Two-dimensional scattergrams of this type will be shown for all sets of V regions subjected to the method of minimum mutations.

Recombination will be detectable by this technique only if it occurs fairly randomly among the family of genes from which the set of proteins under study derive. If it is confined to closely related subsets of these genes, only rarely will the genes expressed in the particular recombinants chosen for study be related in such a way that the apparent genealogical relationship among them will be switched.

The use of genealogical analysis as a critical test of the theories of diversity demands in practice the superior resolving power of the minimum mutations method, in which every possible topology is examined. The limited capacity of that method, however, precludes a simultaneous examination of all the available V regions. Consequently, I have used the method of Fitch and Margoliash in a preliminary survey of the descent of

---

**Figure 6-4.** Detection of recombination by genealogical methods.

A simple recombination event, such as that illustrated here, can switch the genealogical relationship of the recombinant chain to other chains. Provided only that the two genes involved are related to the other proteins in topologically different ways, such a recombination leads to a run of dislocations, reflecting the different relationships of the two parts of the recombinant chain. Detection of such an event does not depend on the total number of genes involved, because it does not require that the genes expressed in the recombinant be the same as those expressed in other proteins studied (a rare occurrence if many genes are involved). Here the two genes expressed in the recombinant–2 and 5–are different from, but related to, the genes expressed in the three other proteins. The sensitivity of the method does depend, however, on the randomness of recombination: if it is supposed that crossing over is limited to narrowly defined, closely related subsets of the genes under study, detectable recombinants would be rare.

If (as supposed in the somatic recombination theory) recombination is frequent and possibly multiple, numerous crossings over which are detectable by the above criterion are expected, each switching the true relationships to a different topological alternative. Many scattered dislocations are therefore expected and no topology is expected to stand out as best. When the excess mutations required by each genealogy for the first half of the chains are plotted against those required for the second half, forming a two-dimensional scattergram, the best topology(ies) for the first half should not be the same as the best for the second half, as illustrated in the figure.

V regions. This survey has defined major families and subgroups of V regions which, even with the uncertainties of this method, are obviously more related to one another than to other V regions. This has been an exercise in confirming the obvious: all the families and subgroups defined in this way are evident by inspection of the sequences and have been accepted by immunologists for several years. These groups of sequences are small enough to be examined by the method of minimum mutations and are few enough in number so that some of the relationships between them can be similarly elucidated.

Figure 6-5 shows an overall descent for many complete V-region sequences reconstructed by the method of Fitch and Margoliash. On the basis of this genealogy, and in accord with the universal opinion of immunologists, three major families of V regions—$\kappa$, $\lambda$, and H—are assumed. As we shall see in Chapter 9, V regions in the same family are related not merely by descent, but also by their location in the genome, and by the C regions with which they are found in the same polypeptide chain; in short, these three families are also physiologically significant units. The next three sections of this chapter will discuss the evolution of each of these families in turn. The results of this study are summarized in the composite genealogy shown in Fig. 6-15. There are only minor differences between the final reconstruction there and the preliminary reconstruction in Fig. 6-5.

## The Evolution of $V_\kappa$ Regions

Figure 6-6 is a reconstruction of the descent of most of the complete or nearly complete $V_\kappa$ regions by the method of Fitch and Margoliash. In this figure, as well as in Fig. 6-7, the subgroups of human (Hood et al., 1967; Hood and Talmage, 1970; Milstein, 1967; Niall and Edman, 1967; Smithies, 1967 a) and mouse (Hood, Potter, and McKean, 1971) $\kappa$ chains that have been previously described are indicated by brackets. These subgroups correspond to genealogical subfamilies in the reconstructed descent. I will assume in what follows that human $V_{\kappa I}$, $V_{\kappa II}$, and $V_{\kappa III}$ are subgroups separate from one another and from the two mouse $V_\kappa$ regions. Moreover, the same three subgroups are evident when the evolutionary relationships are reconstructed separately for positions 1 to 52, 53 to 108, and 1 to 22 (Figs. 6-6 and 6-7). This confirms what several other authors have already pointed out, that the same subgroups of human $V_\kappa$ regions are evident throughout their sequences (Hood and Talmage, 1970; Milstein and Pink,

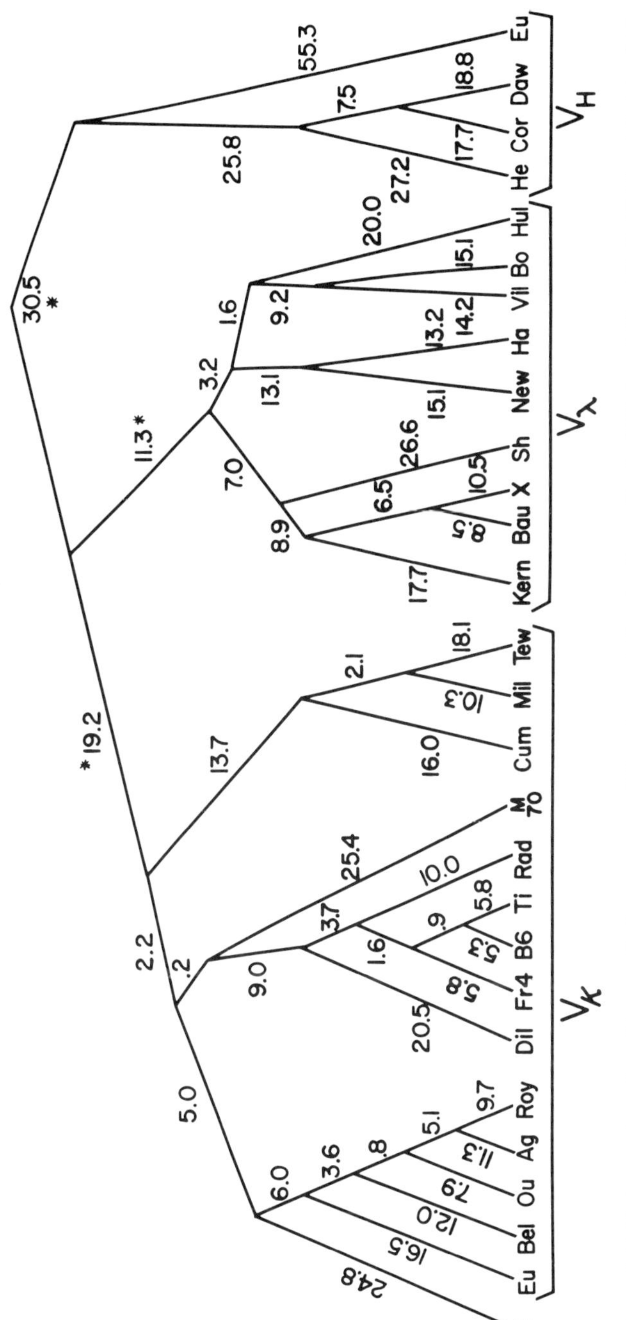


**Figure 6–5.** Best genealogy found in a preliminary survey of V-region evolution by the method of Fitch and Margoliash. The branches marked with an asterisk, which define the $V_\kappa$, $V_\lambda$, and $V_H$ families, will be assumed to be correct in the subsequent examination by the method of minimum mutations.

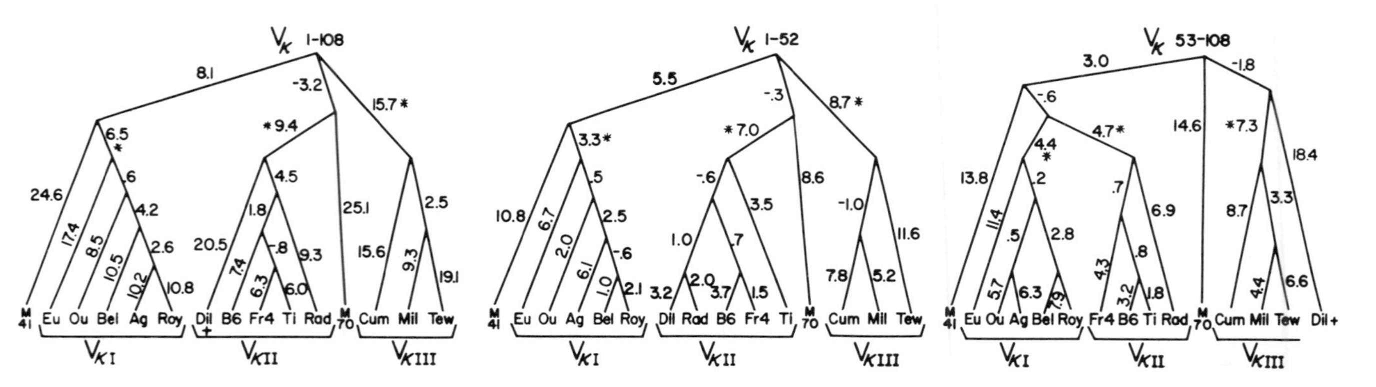


**Figure 6–6.** Best genealogies found in a preliminary survey of $V_\kappa$ evolution by the method of Fitch and Margoliash. The branches marked with an asterisk clearly separate the human $V_\kappa I$, $V_\kappa II$, and $V_\kappa III$ proteins from all others in all three genealogies, indicating that these three subgroups are distinctive throughout the V region and do not recombine. These branches will be assumed to be correct in the subsequent examination by the method of minimum mutations.

+In positions 1 to 52, where its sequence is firmly established, Dil is a typical $V_\kappa II$ protein. There is evidence that the sequence assumed for 53 to 108 (on the basis of preliminary data) is substantially in error. Therefore the placement of this protein at these positions is likely to be incorrect.

1970). These three human subgroups constitute the first solid evidence contrary to the simplest somatic mutation and somatic recombination theories: because each must be coded by a separate gene (or group of related genes), they prove that the germline contains at least three very different $V_\kappa$ genes; because the subgroup relationships persist throughout the V region, none of these chains can be a recombinant between genes from different subgroups.

The genes for $V_{\kappa I}$ and $V_{\kappa II}$ are not allelic to one another, for Milstein, Milstein, and Feinstein (1969) and Smithies, Gibson, and Levanon (1970; Gibson, Levanon, and Smithies, 1971) demonstrated that tryptic peptides characteristic of both are present in each normal individual. $V_{\kappa III}$, a quantitatively minor subgroup of human $V_\kappa$ regions, has not been unambiguously demonstrated in normal light chains and theoretically may be allelic to one of the other two.

I urge the reader not to attach undue importance to these subgroups as physiologically important "units." Similar subgroups, for example, are not seen in mouse $V_\kappa$ or human $V_\lambda$ regions: the obvious groupings are much more numerous in the former (see Fig. 6-7), while there are no such obvious groupings in the latter. It is not true, as we shall see, that each subgroup can be regarded as corresponding to a single germline gene, and if this is to be our criterion we shall become lost in a cumbersome nomenclature of sub-subgroups and sub-sub-subgroups. It is best to regard these groups, when they exist, as accidents of descent reflecting the apparent family relatedness of the sequences chosen for study; we shall then not be surprised by the fact that they do not at all correspond in their general features from one case to the next.

Figure 6-7 is a genealogy reconstructed by the method of Fitch and Margoliash for a large number of partial N-terminal sequences from several species. In this descent the human sequences fall into the above-mentioned three subgroups. The exact relationships shown within these groups, and between these groups and the sequences from other species, are far too subtle to be elucidated by such a limited analysis. However, this genealogy (like the corresponding ones for $V_\lambda$ and $V_H$) firmly establishes two very important points: sequences from other species fall entirely outside the human subgroups (a fact that is conventionally called "species specificity"), and the obvious subgroupings evident in the sequences from these species do not correspond on a one-to-one basis to those of humans. For example, apart from the mouse sequences designated $V_{\kappa I}$ because they are similar to human $V_{\kappa I}$ chains, the other obvious groups of mouse sequences desig-

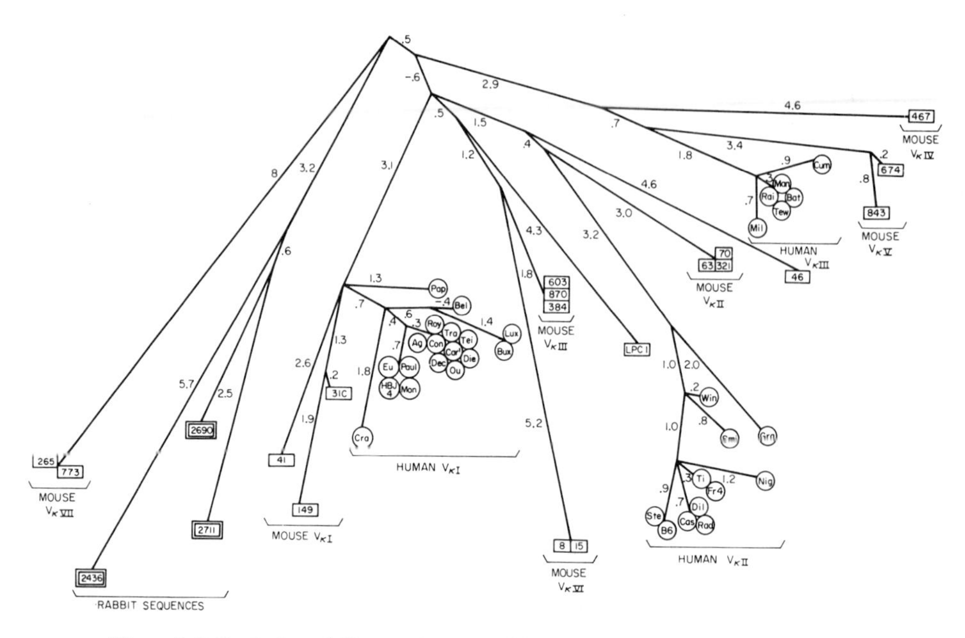


**Figure 6-7.** Evolution of $V_K$, positions 1 to 22; reconstructed by the method of Fitch and Margoliash, and taken from Smith, Hood, and Fitch (1971).

nated in the figure show no simple relation to the human subgroups. These facts strongly suggest (1) that the divergences within the human subgroups are all more recent than the phylogenetic divergence of humans from these other species; and (2) that there are germline genes (corresponding to the subgroups) in one species that are not present (or not expressed) in others, and vice versa. The implication of these conclusions will be discussed in the next chapter.

We shall now turn to an exhaustive analysis by the method of minimum mutations of the divergences within each of the three human $V_K$ subgroups. Figure 6-8 shows the results of three exhaustive analyses of $V_{KI}$ regions. For each analysis the best topology is shown, numbers being assigned to connecting branches by the rules in Chapter 3. Along with each is a scattergram which plots the number of excess mutations required in the first half versus the number required in the second, as explained in Chapter 4. In all these scattergrams, as indeed in every scattergram reported in this book, the best topology(ies) for the first half was (were) the best for the second half as well, within the bounds of uncertainty. Thus the pattern of variation in these chains, as in all other chains I have examined, shows no evidence for recombination.

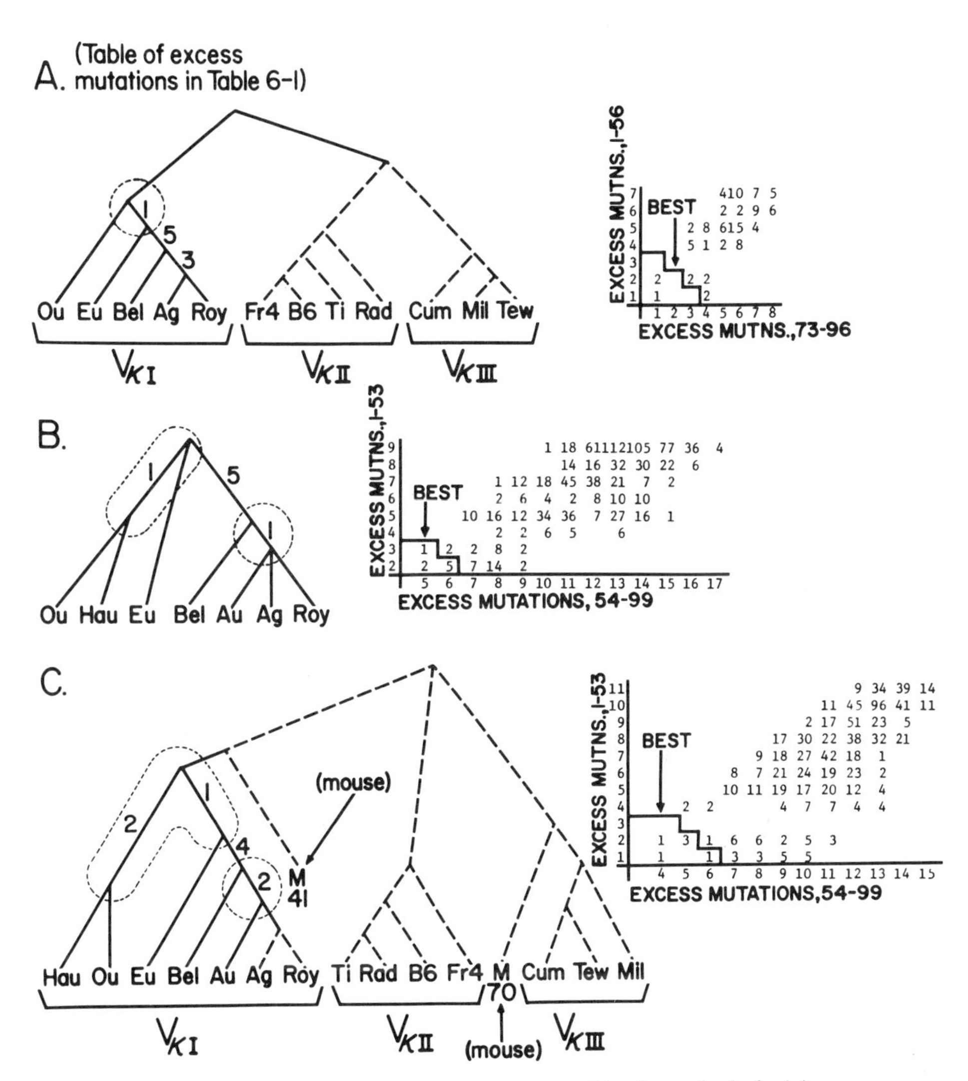


**Figure 6-8.** Evolution of human $V_{\kappa I}$ regions, reconstructed by the method of minimum mutations. Assumed descents* are indicated by dashed lines.

**Justification for assumed descents*: The descents assumed for the human $V_{\kappa II}$ and $V_{\kappa III}$ proteins in Fig. 6-8C are those found in Fig. 6-9A. The descents assumed for these proteins in Fig. 6-8A are different but the differences are unimportant, since the best genealogy in 6-8A conforms to that in 6-8C. The relation between the mouse proteins and the human subgroups assumed in 6-8C is taken from Fig. 6-6 (which places the mouse proteins outside the human subgroups) and Fig. 6-9B (which places MBJ 41 with human $V_{\kappa I}$ and MBJ 70 with human $V_{\kappa III}$).

Table 6-1 shows the excess mutations required at each informative codon position for the same five $V_{\kappa I}$ sequences whose analysis is summarized in Fig. 6-8A. Dislocations, where an alternative genealogy requires fewer excess mutations than the overall best, are entered in large type. This residue-by-residue analysis is designed to detect runs of dislocations indicative of recombination, and as can be seen in the table, a run of four consecutive dislocations is indeed seen at positions 100 to 108 of three genealogies which are otherwise very poor. Evidently at these positions at the very end of $V_{\kappa I}$ regions, the genealogical relationships which hold elsewhere in the chains break down. It seems very unlikely that this scrambling, confined to such a narrow region of chains which otherwise show no evidence for recombination, is directly related to the generation of functional diversity. Rather, its position (at the junction of V and C), and to some extent the pattern of variation itself, suggest some relation to the joining of V and C regions, and discussion will be deferred until Chapter 9. In all the $V_{\kappa}$ data summarized in Figs. 6-8 and 6-9, positions 100 to 108 were ignored. There are only two other positions at which dislocations occur, and no genealogy which contains a dislocation at one of them also contains a dislocation at the other. These dislocations therefore in no way suggest recombination, and the data in Table 6-1 confirm even more strongly the conclusions of Fig. 6-8: that no detectable recombination has occurred among the human $V_{\kappa I}$ proteins.

The data for $V_{\kappa I}$ proteins is strongly suggestive of an ordered genealogy. At the very least, the proteins Ag, Roy, Au, and Bel must be considered separate from the others, and (though less certainly) Bel seems to be separate from Ag, Roy, and Au. Thus there are at least two, and probably three, distinguishable germline $V_{\kappa I}$ genes in humans.

To this multiplicity of germline $V_{\kappa I}$ genes one or two more must be added on the basis of partial sequences which were not included in this analysis. First, HBJ4 is so similar to Eu relative to the other $V_{\kappa I}$ proteins (Hood and Talmage, 1970; Table 2-2A) that a new germline gene, different from that (or those) expressed in Ou and Hau, probably must be admitted for these proteins. Secondly, the presumed identical $V_{\kappa I}$ proteins Dav and Fin, which will be discussed later in this chapter, are certainly sufficiently different from the other $V_{\kappa I}$ proteins to define a new germline gene. The minimum number of human $V_{\kappa I}$ genes therefore stands at three, and if we relax our standard of admission slightly, the number increases to five.

The remaining $V_{\kappa}$ analyses are shown in Fig. 6-9. As shown in Fig. 6-9A, there was no significant evidence for genealogical ordering *within* the

EXCESS MUTATIONS AT FOLLOWING CODON POSITIONS

| TOP. NO. | 111 2235578999000 4800633346046 | TOTAL 1-108 | TOTAL 1-96 |
|---|---|---|---|
| 1 | 1211111 21011 | 13 | 11 |
| 2 | 1211111 21000 | 11 | 11 |
| 3 | 111111 111221 | 14 | 9 |
| 4 | 111111 111221 | 14 | 9 |
| 5 | 1211 11121221 | 16 | 11 |
| 6 | 1211 11121221 | 16 | 11 |
| 7 | 1201121 21011 | 13 | 11 |
| 8 | 1201121 21011 | 13 | 11 |
| 9 | 111111 111210 | 12 | 9 |
| 10 | 11110 111210 | 11 | 8 |
| 11 | 1211 21121211 | 16 | 12 |
| 12 | 1211 21121211 | 16 | 12 |
| 13 | 1211 21 21011 | 13 | 11 |
| 14 | 11 21 1 011 | 8 | 6 |
| 15 | 22111111112210 | 16 | 13 |
| 16 | 22111011112210 | 15 | 12 |
| 17 | 2211121122211 | 19 | 15 |
| 18 | 2211121122211 | 19 | 15 |
| 19 | 1111 1 11221 | 12 | 7 |
| 20 | 11 1 221 | 8 | 3 |
| 21 | 2201111122221 | 18 | 13 |
| 22 | 2201111122121 | 17 | 13 |
| 23 | 22111111122200 | 16 | 14 |
| 24 | 2211111122211 | 18 | 14 |
| 25 | 1211 21 21221 | 16 | 11 |
| 26 | 11 21 1 221 | 11 | 6 |
| 27 | 2201121122221 | 19 | 14 |
| 28 | 2201121122121 | 18 | 14 |
| 29 | 22111111112200 | 15 | 13 |
| 30 | 22111011122210 | 15 | 12 |
| 31 | 11 11 1 011 | 7 | 5 |
| 32 | 11 1 221 | 8 | 3 |
| 33 | 11 11 1 221 | 10 | 5 |
| 34 | 1101111 21011 | 11 | 9 |
| 35 | 1101111121121 | 14 | 10 |
| 36 | 1101111121121 | 14 | 10 |
| 37 | 1 1110 111221 | 12 | 7 |
| 38 | 1111101111221 | 14 | 9 |
| 39 | 1111101111211 | 13 | 9 |
| 40 | 1111 11121221 | 15 | 10 |
| 41 | 1111111121221 | 16 | 11 |
| 42 | 1111111121211 | 15 | 11 |
| 43 | 1 21 1 011 | 7 | 5 |
| BEST | 1 1 211 | 6 | 2 |
| 45 | 1 21 1 211 | 9 | 5 |
| 46 | 121 121 11000 | 10 | 10 |
| 47 | 121 12111200 | 13 | 11 |
| 48 | 121 1111 1200 | 11 | 9 |
| 49 | 111 11 1 1211 | 11 | 7 |
| 50 | 121 121111211 | 15 | 11 |
| 51 | 120 121111211 | 14 | 10 |
| 52 | 121 21111211 | 14 | 10 |
| 53 | 121 1111 1210 | 12 | 9 |
| 54 | 120 121111211 | 14 | 10 |
| 55 | 1201121 21011 | 13 | 11 |
| 56 | 2201111122111 | 16 | 13 |
| 57 | 2201121122111 | 17 | 14 |
| 58 | 1211121 21000 | 12 | 12 |
| 59 | 22111211222 00 | 17 | 15 |
| 60 | 22111111112200 | 15 | 13 |
| 61 | 2211 11122211 | 17 | 13 |
| 62 | 2211 21122211 | 18 | 14 |
| 63 | 1211 21 21211 | 15 | 11 |
| 64 | 2211121112211 | 18 | 14 |
| 65 | 22111111112210 | 16 | 13 |
| 66 | 1211 21 11211 | 14 | 10 |
| 67 | 111110 111221 | 13 | 8 |
| 68 | 22111011112221 | 17 | 12 |
| 69 | 22111011112211 | 16 | 12 |
| 70 | 111111 111221 | 14 | 9 |
| 71 | 2211111122211 | 18 | 14 |
| 72 | 2201111122221 | 18 | 13 |
| 73 | 2211 11122221 | 18 | 13 |
| 74 | 2211 11122211 | 17 | 13 |
| 75 | 1211 11 21221 | 15 | 10 |
| 76 | 2211111 22211 | 17 | 13 |
| 77 | 2201111 22221 | 17 | 12 |
| 78 | 1211 11 21221 | 15 | 10 |
| 79 | 1211 21121221 | 17 | 12 |
| 80 | 2211121122221 | 20 | 15 |
| 81 | 2211111122211 | 18 | 14 |
| 82 | 1211 21121221 | 17 | 12 |
| 83 | 22111111112210 | 16 | 13 |
| 84 | 2201121122221 | 19 | 14 |
| 85 | 2211121112221 | 19 | 14 |
| 86 | 22111111112210 | 16 | 13 |
| 87 | 1211 21 11221 | 15 | 10 |
| 88 | 2211111 22211 | 17 | 13 |
| 89 | 2201121 22221 | 18 | 13 |
| 90 | 1211 21 21221 | 16 | 11 |
| 91 | 1111111 21011 | 12 | 10 |
| 92 | 1 1111 111221 | 13 | 8 |
| 93 | 1111 11121221 | 15 | 10 |
| 94 | 120 121 11011 | 11 | 9 |
| 95 | 111 11 1 1210 | 10 | 7 |
| 96 | 121 21111211 | 14 | 10 |
| 97 | 1211 21 21011 | 13 | 11 |
| 98 | 22111111112210 | 16 | 13 |
| 99 | 2211121122211 | 19 | 15 |
| 100 | 1111 1 11221 | 12 | 7 |
| 101 | 2201111122221 | 18 | 13 |
| 102 | 2211111122211 | 18 | 14 |
| 103 | 1211 21 21221 | 16 | 11 |
| 104 | 2201121122221 | 19 | 14 |
| 105 | 22111111112210 | 16 | 13 |

**Table 6-1.** Excess mutations for Fig. 6-8A ($V_{\kappa I}$).

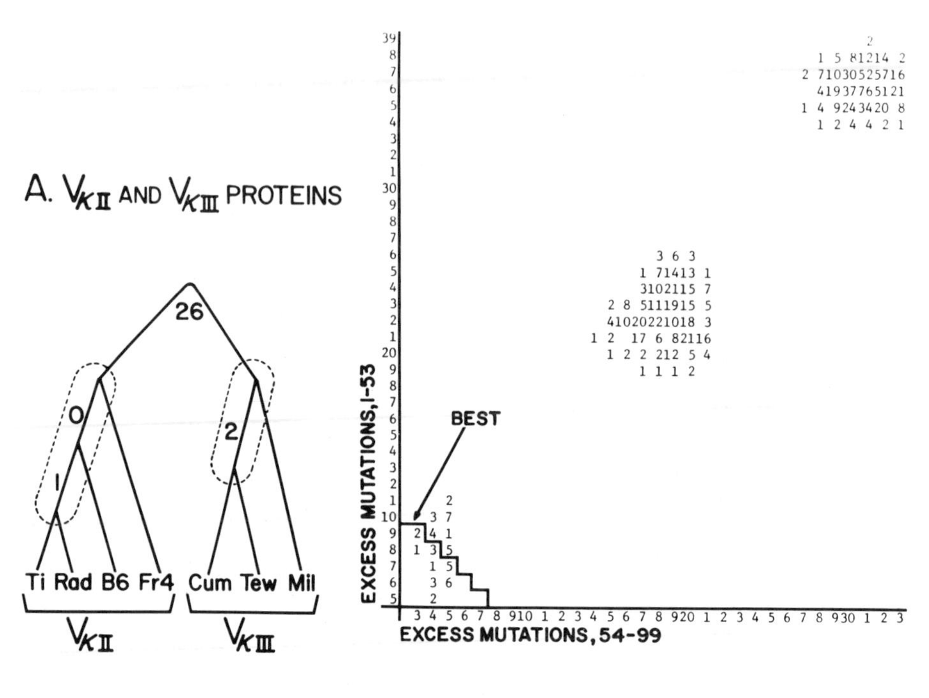


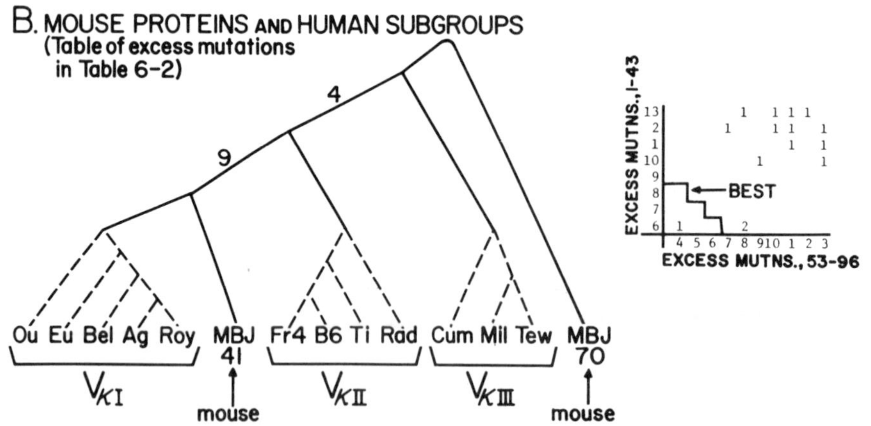


**Figure 6–9.** Evolution of human $V_{\kappa II}$ and $V_{\kappa III}$ regions, and the relation of mouse $V_{\kappa}$ sequences to human $V_{\kappa}$ subgroups; reconstructed by the method of minimum mutations. Assumed descents* are indicated by dashed lines.

**Justification for assumed descents*: The descent for $V_{\kappa I}$ in Fig. 6–9B is that found in Figs. 6–8A, 6–8B, and 6–8C. The $V_{\kappa II}$ and $V_{\kappa III}$ descents assumed in Fig. 6–9B are different from those found in 6–9A, but the differences would not significantly affect the results. In accord with Fig. 6–6, the mouse proteins were assumed in 6–9B to fall outside the human subgroups.

human $V_{\kappa II}$ and $V_{\kappa III}$ subgroups. This lack of order did not result from a contradictory pattern of mutations, with many dislocations, but rather from a paucity of positions which were informative about the intrasubgroup divergences. As discussed above, this paucity implies a corresponding paucity of mutations during the germline divergences of these proteins and thus constitutes weak evidence against a germline origin of their differences.

The relation of the two mouse sequences to the human subgroups is shown in Fig. 6-9B, and the table of excess mutations for this analysis is shown in Table 6-2. MBJ41 is topologically related to $V_{\kappa I}$, while MBJ70 is topologically related to $V_{\kappa III}$. There is no evidence for recombination involving these chains.

| TOP. NO. | EXCESS MUTATIONS AT CODONS 11 1112222233455567778999900 34935901292633450247513608 | TOTAL 1-108 | TOTAL 1-96 |
|---|---|---|---|
| 1 | 211101111 0011   1011113000 | 19 | 19 |
| 2 | 211111101111 01111 11   121 | 22 | 19 |
| 3 | 20 111011 11111111111113121 | 26 | 23 |
| 4 | 211111111110011111011113000 | 23 | 23 |
| BEST | 1  1 11   11 1    1  1   121 | 13 | 10 |
| 6 | 2111110111111111111111013121 | 28 | 25 |
| 7 | 2111111111100 1111011113000 | 22 | 22 |
| 8 | 0  1 11   1111     11 112121 | 17 | 14 |
| 9 | 2111111011111011111110 3121 | 26 | 23 |
| 10 | 211111111111 11111 11   121 | 24 | 21 |
| 11 | 20 111111 11111111111113121 | 27 | 24 |
| 12 | 211101111 1111   1111013111 | 24 | 22 |
| 13 | 211111011111 11111111113121 | 28 | 25 |
| 14 | 211111101111101111111 3121 | 27 | 24 |
| 15 | 1  0 11   1111     11 112111 | 16 | 14 |

Table 6-2. Excess mutations for Fig. 6-9B (human and mouse $V_{\kappa}$).

## The Evolution of $V_\lambda$ Regions

The λ chains were treated in a manner similar to the κ chains. In Fig. 6–10 is the single best descent consistently found by the method of Fitch and Margoliash for positions 1 to 112, 1 to 34, 35 to 72, and 73 to 112 of a number of complete $V_\lambda$ regions; the conformity of all four genealogies to one another strongly suggests that little if any recombination has occurred in these chains. In Fig. 6–11 is a genealogy reconstructed by the same method for a number of short N-terminal sequences; as for the κ chains, the sequences from other species did not mix with their human counterparts, confirming the "species specificity" already noted in the κ chains. It should be noted that there are in the human $V_\lambda$ sequences no strikingly obvious subgroups in the first twenty positions. The previously defined subgroups (Hood and Ein, 1968 b; Hood and Talmage, 1970; Langer, Steinmetz-Kayne, and Hilschmann, 1968) are indicated in Figs. 6–10 and 6–11, but they do not correspond exactly to genealogical families. In this region neither the exact structure of the genealogy nor the published subgroup assignments should be taken very seriously. Even in the genealogy for the complete sequences (Fig. 6–10), the divergences

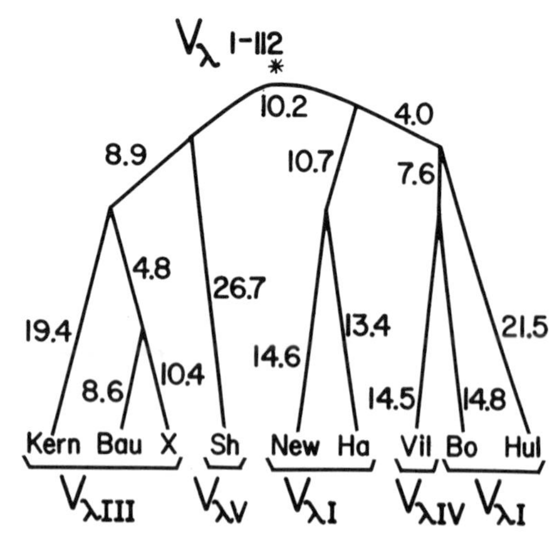


**Figure 6–10.** Best genealogy found in a preliminary survey of $V_\lambda$ evolution by the method of Fitch and Margoliash. Genealogies reconstructed separately for positions 1 to 34, 35 to 72, and 73 to 112 of these chains were identical to that shown here, indicating that these sequences are not recombinants. The branch marked with an asterisk will be assumed to be correct in the subsequent examination of $V_\lambda$ evolution by the method of minimum mutations.

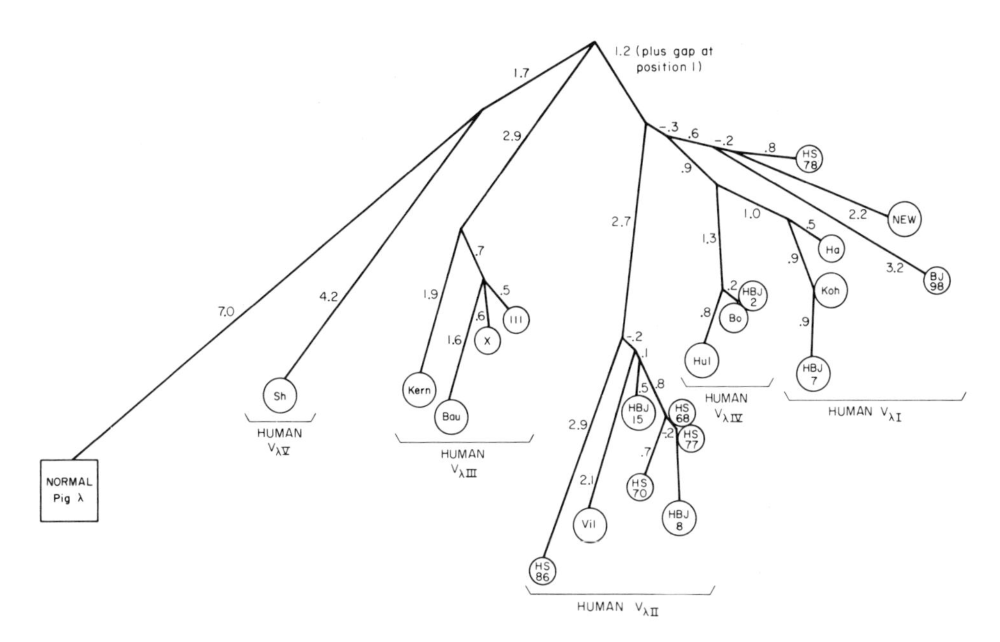


**Figure 6–11.** Evolution of $V_\lambda$, positions 1 to 20; reconstructed by the method of Fitch and Margoliash, and taken from Smith, Hood, and Fitch (1971).

appear to be more evenly distributed in the evolutionary history of these proteins and not (as in the κ chains) confined for the most part to relatively recent divergences within well-defined major subgroups.

In analyzing the descent of $V_\lambda$ regions more closely by the method of minimum mutations, I assumed that the starred branch in Fig. 6–10 divides the sequences into two nonrecombining sets which could be treated separately. This assumption was partially tested and confirmed in several of these analyses.

The $V_\lambda$ analyses by the method of minimum mutations are shown in Fig. 6–12. Tables 6–3 and 6–4 show the tallies of excess mutations by codon position for the analyses in Figs. 6–2A and 6–2C respectively. None of the scattergrams showed any evidence of recombination, nor do the scattered patterns of dislocations in Tables 6–3 and 6–4 suggest any recombination events, even at the end of the $V_\lambda$ regions. These analyses point to a highly ordered $V_\lambda$ genealogy, and from this we may deduce a minimum of four different human $V_\lambda$ genes. The divergences distinguishing the sequences X, Bau, and Kern (which are classified together as subgroup $V_{\lambda III}$) could not be ordered with any certainty. This lack of order is not attributable to a lack of informative positions, but rather to the fact that the testimonies of

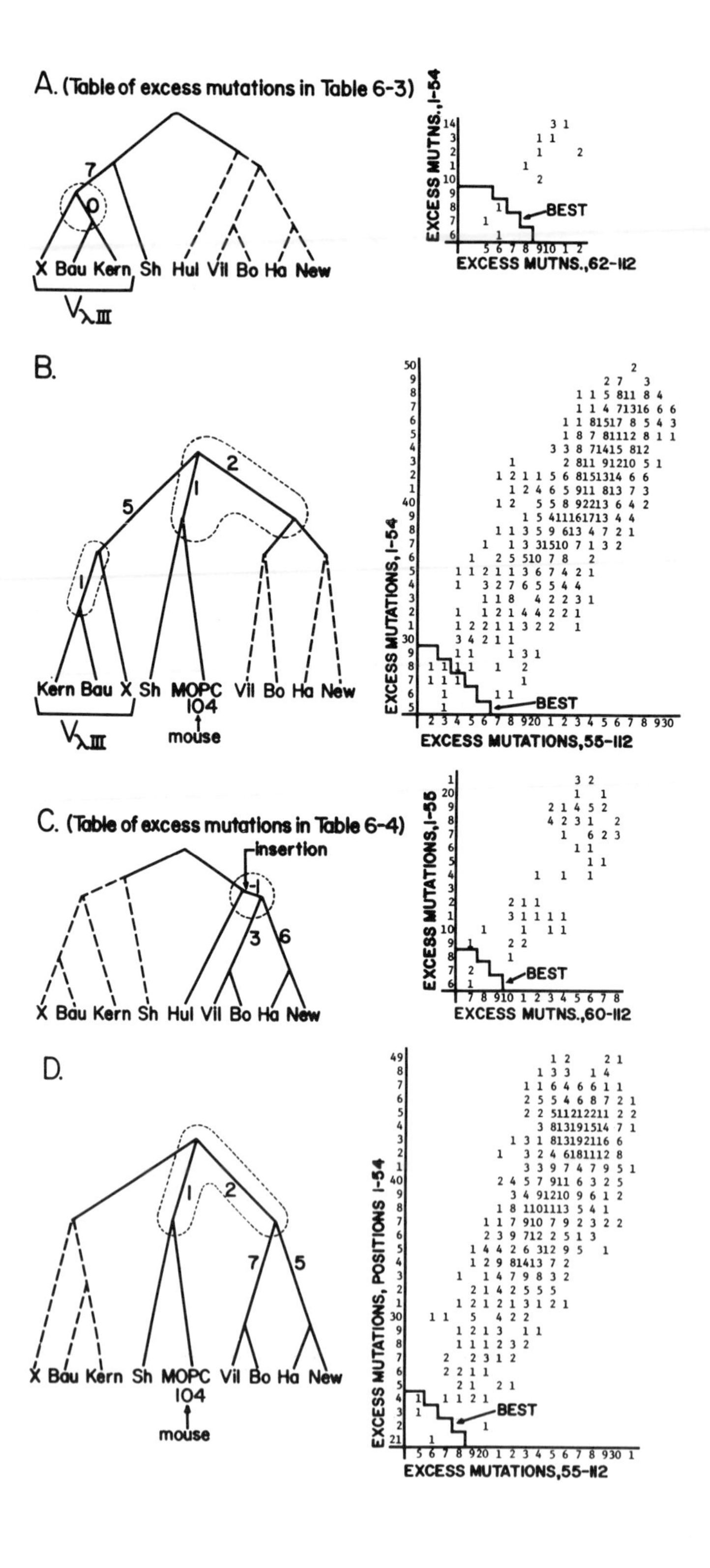

A. (Table of excess mutations in Table 6-3)
X Bau Kern Sh Hul Vil Bo Ha New
BEST
EXCESS MUTNS.,1-54
EXCESS MUTNS.,62-112
B.
Kern Bau X Sh MOPC 104 Vil Bo Ha New
mouse
BEST
EXCESS MUTATIONS,1-54
EXCESS MUTATIONS,55-112
C. (Table of excess mutations in Table 6-4)
insertion
X Bau Kern Sh Hul Vil Bo Ha New
BEST
EXCESS MUTATIONS,1-55
EXCESS MUTNS.,60-112
D.
X Bau Kern Sh MOPC 104 Vil Bo Ha New
mouse
BEST
EXCESS MUTATIONS, POSITIONS 1-54
EXCESS MUTATIONS,55-112

| | EXCESS MUTNS. AT CODONS | TOTAL | |
|---|---|---|---|
| | (*gap at 98-99) 11 | M | G |
| TOP. | 123333444555666777888999 01 | U | A |
| NO. | 2386234615724281403267*12 | T | P |
| 1 | 1012 1011111011 11211 * 1 | 19 | 1 |
| 2 | 11121111011211111112101*10 | 24 | 1 |
| 3 | 11111 1011021101101 11*11 | 19 | 1 |
| 4 | 101211011111201111112111*11 | 24 | 1 |
| 5 | 1112 1111111111111112101* 1 | 23 | 1 |
| 6 | 1 01 101 011 11 00 11 11 | 12 | 0 |
| 7 | 101211011111201111112111*11 | 24 | 1 |
| 8 | 1112 11111011101102111* 1 | 21 | 1 |
| 9 | 1 11 110 111 11 11 01 10 | 14 | 0 |
| 10 | 101211111111211111112101*11 | 25 | 1 |
| 11 | 11121111111021101102111*11 | 24 | 1 |
| BEST | 1 1 111 1 1 1 11 1 1 | 12 | 0 |
| 13 | 10111 10110211111101 11*11 | 19 | 1 |
| 14 | 11121111011211111112101*10 | 24 | 1 |
| 15 | 1112 1111111111 11211 * 1 | 22 | 1 |

**Table 6-3.** Excess mutations for Fig. 6-12A ($V_\lambda$ sequences X, Bau, Kern, Sh).

**Figure 6-12.** Evolution of $V_\lambda$, reconstructed by the method of minimum mutations. Assumed descents* are indicated by dashed lines.

**Justification for assumed descents*: The descent assumed for Hul, Vil, Bo, Ha, and New in Figs. 6-12A and 6-12B was that found to be best in Figs. 6-12C and 6-12D. The descent assumed for X, Bau, and Kern in Fig. 6-12D was that found to be best in Figs. 6-12A and 6-12B. The descent assumed for X, Bau, Kern, and Sh in Fig. 6-12C differed only in unimportant respects from that found to be best in 6-12A and 6-12B, and the best topology found in 6-12C conformed to that found in 6-12D.

```
          EXCESS MUTNS. AT CODONS            TOTAL
      -------------------------------------  -----
           (*gap at 27-29)                1   M G
TOP.   1122 2333344455556777899999991   U A
NO.   30735*8012614913450126113567892   T P
----  -------------------------------  --- -
   1  10112*120 11 1011111212121 20 1   28 1
   2  10113*12011111011111212121130 1   33 1
   3  1111 2*1211111111101121012012110   31 1
   4  11113*1211111111101121112013110   34 1
   5  111 1*  1 011011111101211112111   25 1
   7  10112*120 11 1011121212121 21 1   30 1
   8  10113*12011111011121212121131 1   35 1
   9  11112*1211111111102121012112111   34 1
  10  11113*1211111111102121112113011   36 1
  11  111 1*  1 111011111101211112111   26 1
  13   0 12* 20 1  101 12 1 1 21 11 1   19 1
  14   0  2* 20 1 1100 12 1 1 11111 1   18 1
  16  11113*1211011112112111113113011   37 1
  17  11113*1211111112111111113013110   36 1
  18  11112*1111111112110111113012110   32 1
  19   1 12* 21 1 1111 02 1 0 21111 1   22 1
  20   1  2* 21 1 1100 02 1 1 11111 1   19 1
  24  11112*1111111112110111213112011   34 1
  25   1   *  1 1 1011 11 0 1 111 1 1   14 1
  26   1   *  1 1 1100 11 1 1 011 1 1   13 1
  27  11113*1211111112111121113013110   37 1
  28  11113*1211111112111121113013110   37 1
  31   0  2* 201111100 11 212 11121 1   23 1
  32   1  2* 211111100 01 211 11121 1   23 1
  33   1  1*  11111100 11 112 01121 1   20 1
  34  10113*12011111011111212121131 1   34 1
  36  11113*1211111111111121112013110   35 1
  37  11113*1211111111101121112113011   35 1
  38  11113*1211011111111111212113011   35 1
  41  11112*1111111111110121112012110   31 1
  42  11112*1111111111110121212112011   33 1
  43   0  2* 201111100 12 212 11121 1   24 1
  44   1  2* 211111100 02 211 11121 1   24 1
  45   1  1*  11111100 11 112 01121 1   20 1
  46  10113*12011111011121212121131 1   35 1
  55  10113*12011111011121111121131 1   33 1
  57  11113*1211111112111111113013110   36 1
  58  10113*12011111011121111121131 1   33 1
```

(continued)

```
          EXCESS MUTNS. AT CODONS          TOTAL
       (*gap at 27-29)              1      M G
TOP.    1122 23333444555556777899999991     U A
NO.    30735*801261491345012611356789 2     T P
 61    11112* 11 0110121121011131121 11    31 1
 62    11112* 11 1110121111011131121 11    31 1
 63     1 12* 11 1 1011 12 0 1 21111 1     21 1
 64    11112*121111111121011110130121 10   32 1
 65    11112*121111111121021110131121 11   35 1
 66     1 12* 21 1 1111 02 1 0 21111 1     22 1
 67    11113*121 1110111021111121130 11    33 1
 68    11113*121 01101211211121311301 1    36 1
 69    11113*121 11101211111112131130 11   36 1
 70    11113*121 1110111021111121131 11    34 1
 73    11112* 11 0110121121012131121 11    32 1
 74    11112* 11 1110121111012131121 11    32 1
 75     1 12* 11 1 1011 12 0 1 21111 1     21 1
 76    11112*121 11 0121111112131 20 1     31 1
 77    11112*121 11 0121121112131 21 1     33 1
 78     1 12* 21 1  011 12 1 1 21 11 1     21 1
 79    111       1111 111110111111111 1111 23 0
 80    11112 111111 11211012111130 2110    31 0
 81    11112 111111 112110121111 31 2011   32 0
 82    111       1111 11111111111111 1111  24 0
 84    11112 121111 1121111211131 2111     35 0
 85    11112 121111 112101121013 0 2110    31 0
 86    11112 121111 11210112101 31 2111    33 0
 87     1  1  21 1  111 01 1 0 21   1 1    16 0
 88    11112 121 11 1121111211131 20 1     32 0
 89    11112 121 11 1121111211131 21 1     33 0
 90     1  1  21 1  111 11 1 1 21   1 1    18 0
 91    10113*120111110111112121211 30 1    33 1
 92    11113*121111111111011211120131 10   34 1
 94    10113*1201111101111212121211 31 1   35 1
 97     0 12* 20 1 1101 12 1 1 21111 1     21 1
 99    11113*121111111121111111130131 10   36 1
100     1 12* 21 1 1011 02 1 1 21111 1     22 1
101    11113*121 0110121121 1213113111     36 1
102    11113*121 1110121111 12131130 11    35 1
BEST    1       1 1  111 11 1 1 11   1 1   14 0
104    11112 121111 11211112111 30 2110    33 0
105    11112 121111 1121111211131 2011     34 0
```

**Table 6-4.** Excess mutations for Fig. 6-12C ($V_\lambda$ sequences Hul, Vil, Bo, Ha, New).

many informative positions disagree on their descent; this suggests a high incidence of parallel and back-mutations, and no conclusion is possible as to the contribution of germline mutations to their differences.

## The Evolution of $V_H$ Regions

A genealogy reconstructed by the method of Fitch and Margoliash for many short N-terminal sequences is shown in Fig. 6-13. As in the case of $V_\kappa$, the human sequences fall into three major subgroups, which have been previously described in the literature (Cunningham et al., 1969; Kohler

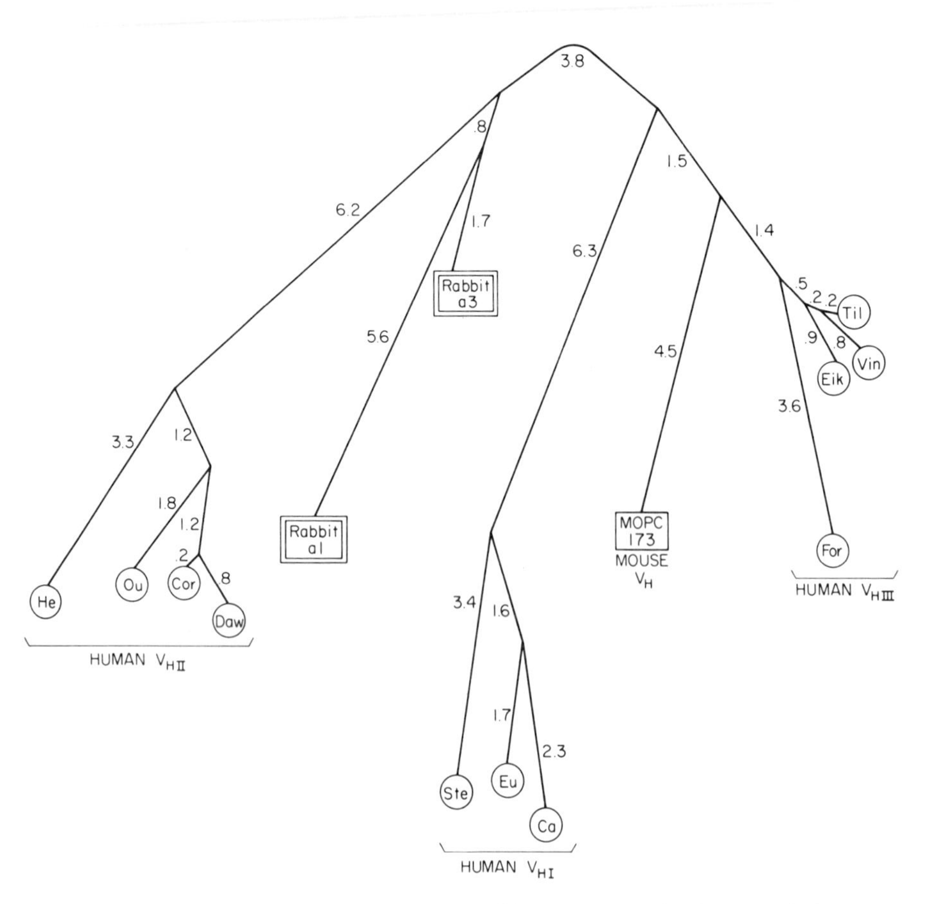


**Figure 6-13.** Evolution of $V_H$, positions 1 to 34; reconstructed by the method of Fitch and Margoliash, and taken from Smith, Hood, and Fitch (1971).

et al., 1970a; Press and Hogg, 1969; Wang et al., 1970a), and the three nonhuman sequences fall outside the human subgroups.

The six complete $V_H$ sequences were subjected to an analysis by the method of minimum mutations. The results are shown in Fig. 6-14A, and a table of the excess mutations at each informative nucleotide position is given in Table 6-5. The scattergram shows no evidence for recombination, and while the analysis shows a clear genealogical separation of the four $V_{HII}$ proteins from the other two, there is no clear evidence for any order of the divergences within the $V_{HII}$ subgroup. As in the case of $V_{\lambda III}$, this lack of order is attributable to disagreement among an abundance of informative positions.

In Table 6-5 there is a run of four consecutive dislocations at positions 50.2 to 59.1 in three of the genealogies which were otherwise poor. By itself this could not be taken as significant evidence for recombination, in view of the large overall incidence of dislocations (thus, presumably, parallel and back-mutations) evident in Table 6-5. However, one of these three is also supported by the pattern of tryptophan insertions at position 52a (see Table 2-4A), which falls within this short segment. The alternative topology suggested by the pattern of mutations and gaps in this region is shown in Fig. 6-14B. There is another adjacent position (49a) in which tryptophan insertions are apparent; these insertions are consistent with neither the overall best topology (Fig. 6-14A) nor with the alternative in Fig. 6-14B. Altogether, I am suspicious of these tryptophan insertions; until their locations are confirmed, the possibility must be kept in mind that these residues (which are notoriously conducive to sequencing errors) have been incorrectly placed in the published sequences. It is also conceivable that the alignments chosen for the sequences in this region are not realistic. Apart from the scrambling at the end of $V_\kappa$, this is the only case for recombination that can be made in the pattern of mutations in immunoglobulin chains, and it is a marginal one.

The evolution of V regions is summarized in the composite genealogy in Fig. 6-15. The different human germline genes which are inferrable from the sequence evidence are indicated by brackets (the less certain ones within the $V_{\kappa I}$ subgroup by dashed brackets). This analysis is entirely consistent with the expectations of the germline theory and introduces complications of differing seriousness for the somatic theories.

The failure to find any significant evidence for recombination—except the presumably irrelevant case at the end of $V_\kappa$ and the doubtful case in

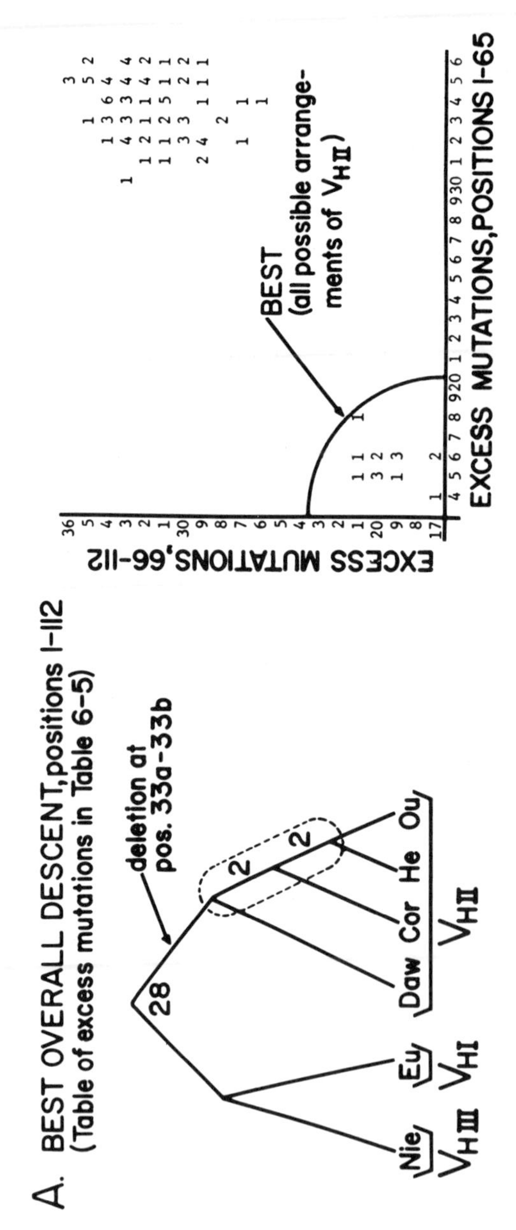


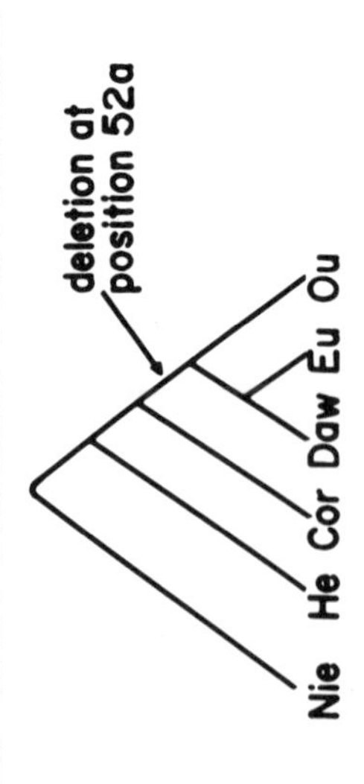


Figure 6–14. Evolution of $V_H$, reconstructed by the method of minimum mutations.

the middle of $V_H$—seems to preclude the simplest version of the somatic recombination theory. That theory had already to be radically modified in an ad hoc manner to explain why no recombinants have been observed between major subgroups of human V regions (Gally and Edelman, 1970); to this must now be added the further complication that no recombinants have been observed within the human $V_{\kappa I}$ subgroup, in which (because of the clear genealogical separation of these proteins) recombination could have been detected. The somatic recombination theory thus must restrict the putative crossing-over events to ever-narrowing subgroups of genes, imposing an entirely gratuitous constraint on the system to generate diversity. In support of this conclusion, Smithies, Gibson, and Levanon (1970) found in normal human $\kappa$ chains none of the peptides that would be expected if recombination occurs detectably between the arginine residues at positions 18 and 24 of the human $V_\kappa$ genes. Recombination, whether germline or somatic, is expected to occur occasionally in any theory, of course. The sum of the evidence is that it is at most an occasional event, and far from the frequent one expected if the somatic recombination theory is correct. I feel justified, therefore, in dismissing this theory for the remainder of this book.

The restrictions this genealogical evidence imposes on the somatic mutation theory, however, are far less severe. The maximum total number of human $V_\kappa$ genes that can be inferred from it, for example, is only seven, far from the 425 human $V_\kappa$ genes which must be postulated if the germline theory is correct. Thus even if proponents of somatic mutation admit all the foregoing conclusions, they are not forced to accommodate an implausibly large number of germline genes. Indeed, once it had been shown by Dreyer and his associates (Dreyer, Gray, and Hood, 1967; Hood et al., 1967) that the number of human germline $V_\kappa$ genes is more than one, the genealogical evidence had already extracted its maximum concession from the somatic mutation hypothesis, for in the balance of judgment there is surely a negligible difference between two and seven genes.

Moreover, all the foregoing conclusions need not be accepted without reservation. It is conceivable that linked, parallel somatic mutations could lead to a pattern of variation which mimics that expected from multiple different germline genes. Such a pattern might result, for instance, from a special somatic hypermutation mechanism from which only a few patterns could result, or from a special scheme of somatic selection. These reservations do not seem very important, however, for

```
                  EXCESS MUTATIONS AT FOLLOWING CODON & NUCLEOTIDE POSITIONS
                                                                          11       T
                                                                     1111001111111 0
            1111222222333333344555555666666777777788888888899900002200000011 T
TOPOL. 33445915571448912233457040022912256601234488002567793990012aa4456922 A
   NO. 12121111211121121212111121212111211211211212122211222121222121322112 L
     1 111 1111111 111121121110111110111110100122111111  121111111111111112111 66
     2 11111111111111111210211111111111111111101122111111  1211111 111111112111 70
     3 111 1111     111111002210121111111121112011100111 1  11 211011111201111 2011 62
     4 1111111111111111100221012110011111112011111111111111121101111111101120 11 65
     5 111111111111111  212 11121 1101 1111211111221111   11   1100 121 111 1    59
     6 111111111111111  212111121 1101 2101211112211111  1111110011111 11111 1 65
     7 111 1111111 111121121110110010111110110122111111  12111101101111112111 63
     8 1111111111111111111211111001012111111012211111 1  121110110111111 2111 67
     9 111 1111     111111101121012111111121112011000111 1  11 211111102011112111 62
    10 11111111111111111101211121111111111112011011111111111211101102011112111 68
    11 111111111111111  211  11121 1111 1111211111221111   11   1111 121 111 1   61
    12 111111111111111  1111111211111  11112111122111111  1111011 121 11111 1 66
    13                  12111 1 0111110 11 1 0110111            1  1111111111 1111 35
    14                  11111 1 1011111112 1 1111111            1  111111 11 1111 37
    15 11111111111111111211111122 110112101211101111111111111100101011111111 65
    16 11111111111111111111111122 110111111211101111111111111100102011111111 65
    17 1111111111111111111110122100111111121111111111111111011 1111101101 1 64
    18 1111111111111111111110122111111111121111111111111111011 121 1111011 67
    19                  110021 0 111111112 1 10111 00           1  1111120111 1111 35
    20                  110021 1 1011111112 1 1011111           1  111111 11 1111 36
    21 11111111111111111212111112 0000121011110122111111111111100111111111111 65
    22 1111111111111111112111112 0000121111110122111111111111100111111111111 65
    23 1111111111111111121011111211111011111011221111111111111111 1111111111 68
    24 11111111111111111111111112111110111110112211111111111111011 121 1111111 67
BEST                    212   1 11 1111 1 1 1111111                111 121 11  1   31
    26                  1121 1 1011111 2 1 1111111             1  111111 11 11 1 35
    27 11111111111111111122101121001011111111011111111111211010111111012011 64
    28 11111111111111111122101121001012111111011111111111211010111111112011 66
    29 1111111111111111121021111211111111111101011111111112111111 01011112111 67
    30 111111111111111111121111211111111111101011111111112111011020111121 11 68
    31 11111111111111111112111101111112111111112211111   12111111111 1112111 70
    32 111111111111111110022111201111112111201111111111   12111111111 1112111 68
    33 111111111111111   12111120111111 21112111122111111   11111111111 11111 1 67
    34 11111111111111111112111111111121111111122111111 12111101111 1112111 71
    35 11111111111111111121111121 110112111211112211111111111100111 1111111 69
    36 111111111111111110122101211111112111211111111111111121101011111 1112011 69
    37 11111111111111111002211121111111111112011111111111121110110 21 1112111 69
    38 11111111111111111121111121 11011111121111221111111111100101021 1111111 68
    39 11111111111111111112111111111111111111101122111111112111011021 1112111 70
    40 111111111111111   1211112111111  1111211112211111  111110111121  11111 1 68
    41 111111111111111110122101211111111111211111111111112110111121 1112011 70
    42 1111111111111111111121111111111111111101122111111112110111121 1112111 71
    43 11111111111111111112111101111112111111112211111   1211111110 1 1112111 69
    44 1111111111111111110121112011111121112011111111111   12111111111 1112111 68
    45 111111111111111  11111112011111  2111211112211111   111111111111 11111 1 66
    46 1111111111111111121021111111111111111111112211111  1211111 1011111 2111 70
    47 11111111111111111210111121111111111112111122111111111111111 1111111111 71
    48 111111111111111112102111211111111111121110111111111121111 1 0101111 2111 69
    49 1111111111111111110121112110011111111201111111111111211111111111101201 1 67
    50 111111111111111111111111211001111111211112211111111111111 1111101101 1 67
    51 11111111111111111112111111001 11111111012211111111111211101101110120 11 65
    52 111111111111111  2111111211111  2101211112211111  11111110111 1  11111 1 67
```

(continued)

```
                         EXCESS MUTATIONS AT FOLLOWING CODON & NUCLEOTIDE POSITIONS
                                                                               11         T
                                                                      1111001111111       0
                 11112222233333333445555566666677777777888888888999000022000001 1         T
       TOPOL. 33445915571448912233457040022912256601234488002567793990012aa4456922         A
          NO. 12121111211121121212111121212111211211211212122211222121222121322112         L
   53 111111111111111121121112111111121012111011111111111121110101011112111  70
   54 111111111111111121121111110010121011110122111111111121111011011111 2111  68
   55 1111111111111111111111111111111211111111111111 1 1111110 1111111111  65
   56 1111111111111111111111122 110112111211111111111 11111100 1111111111  66
   57 1111111111111111111110122111111211121111111111 11111010 11111111011  66
   58 11111111111111112101111111111111111111111111 1 1111111 11111111111  65
   59 11111111111111112101111221111111112111111111 111111111 11111111111  69
   60 11111111111111112101111221111111112111011111 111111111 01011111111  66
   61 1111111111111111211 11122 1101111111211111111 11  1100 1211111 11  61
   62 1111111111111111211 11122 1111111112111111111 11  1111 1211111 11  64
   63                 1211  1 11 111111 1 1111111        111 121111  11  34
   64 111 1111  1111111111012211111211121110011111 11 111011 12011111011  61
   65 111 1111  1111111111012211111211121110001111 11 111111 0201111111  61
   66                 11111 0 111111112 1 111100        1  111 120111 1111  35
   67 11111111111111111021111111111111111110111111111111111110110211001111  63
   68 1111111111111111112111112 1101111111111112211111111111001021100 1111  65
   69 111111111111111111111111211111101111101122111111111111011021100 1111  65
   70 1111111111111111102111111100111111110111111111111111111111111011 1  61
   71 111111111111111111111111210011101111111122111111111111 1111001101  63
   72 11111111111111111211111210010111111101221111111111110111111001101  63
   73 1111111111111111212 11112 1101111111111221111  11  1100 12110011 11  60
   74 111111111111111121 1 11112 111101111111111221111  11  1111 12110011 11  61
   75                 1212  1 11 111111 1 1111111        111 121100  11  33
   76 111 1111111 11112111110211101011101001221111111 1111111111111001111  60
   77 111 1111111 111121211110210010111110110122111111 11111101111100 1111  60
   78                 12121 1 01111101 1 1 0110111       1  1111111100 111  33
   79 111 11111111111 11211111111111 1101111111111111 1111101111 21 11111 1  62
   80 111 11111111111111221011211111111101111111111111111112110111121 1112 11  68
   81 111 11111111111111121111211111111011101111111111111121101111 21 1112111  68
   82 111 11111111111 21211111111111 2101111111111111 11111101111 1 11111 1  63
   83 111 1111111111111211211112111111121011111011111111121111010101111 2111  68
   84 111 111111111111212211112100101210111101111111111112111101111111 12111  68
   85 111 1111  111111112210112111111210111111001111 11 21101111201111 2011  63
   86 111 1111  1111111112101121111112101111100011111 11 21111110201111 2111  61
   87                 11121 0 111111112 0 111100        1  1111120111 1111  36
   88 111 1111111 111121121110211101110101001111111111 121111111111112111  64
   89 111 1111111 1111212211102100101110101101111111111 12111101111111 2111  63
   90                 12121 1 0111101 1 0 0110111       1  11111 1111 1111  35
   91 11111111111111111111211111111111111111101122111111 121111111 1 1112111  70
   92 1111111111111111100221012111111112111 2011111111111111211011 1121 1112011  69
   93 1111111111111111 1121111121 1101 21112111122111111 11111001121 11111 1  66
   94 111111111111111121121111110010121111111012211111 12111101101111112111  68
   95 11111111111111111010121112111111121112011011111111111121111101011112111  69
   96 1111111111111111 211111121111111 111121111221111 11111111 111 11111 1  67
   97                 12111 1 111111112 1 1111111       1  111 111111 1111  39
   98 111111111111111121111122 1101121112111011111 11111100 0201111111  65
   99 11111111111111111111101221111111111121111111111 11111011 12111111011  67
  100                 11021 1 111111111 1 1011111       1  111112110 0 111  35
  101 11111111111111121211111 2 00001111111101221111111111111100111 1100 1111  62
  102 111111111111111121 111112111111011111011221111111111111111 12110 01111  67
  103                 2121 1 1111111 2 0 1111111        1  1111121 11 11 1  37
  104 111 111111111111112210112 1 01012101111011111111111112110101111111112011  64
  105 111 11111111111121121111211111111011101011111111111121111102011112111  68
```

Table 6-5. Excess mutations for Fig. 6-14 ($V_H$).

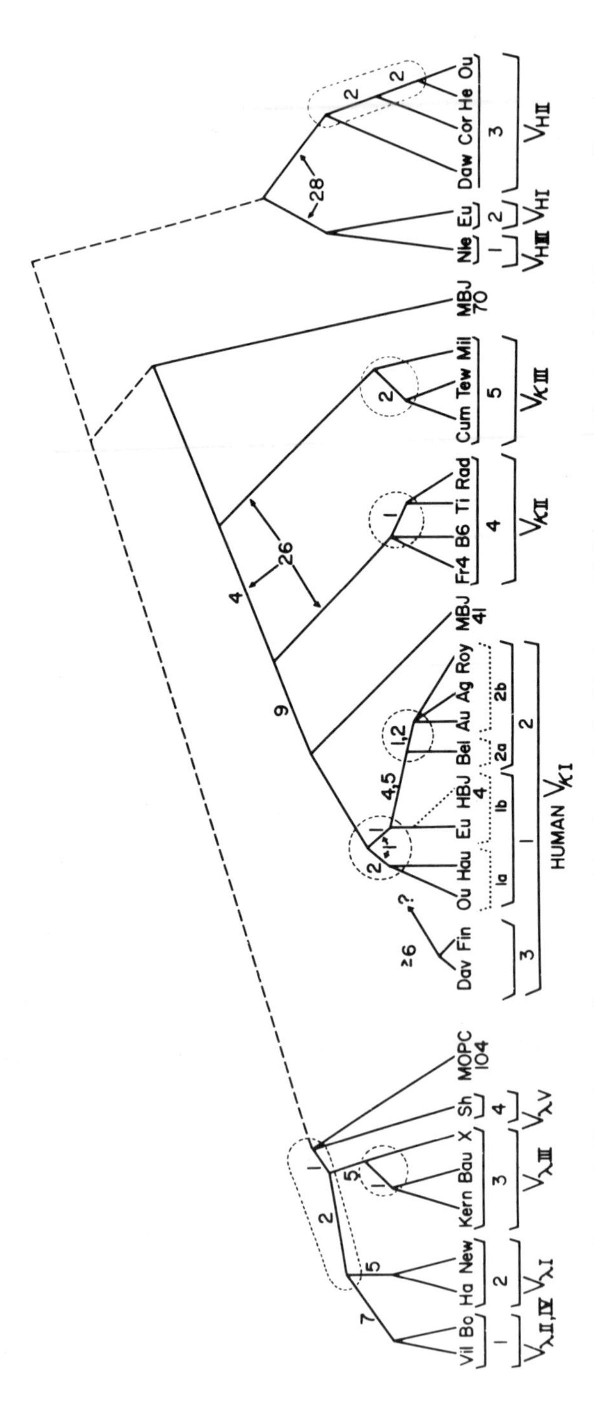


**Figure 6–15.** Composite V-region genealogy, summarizing the results in this chapter. Nodes corresponding to minimal human germline genes are indicated by numbered brackets (the less certain ones, by dotted brackets). The relationship among the $V_\kappa$, $V_\lambda$, and $V_H$ families is taken from Dayhoff (1969).

they involve special assumptions which are at least as gratuitous as the admission of a few more genes into the germline complement.

V-region sequence identities bear critically on the choice between the germline and somatic mutation theories. The basic premise by which these identities are interpreted is the same genealogical reasoning used in the foregoing analysis of V-region evolution. If two sequences from different individuals are identical, I assume that their germline genes were also identical, for otherwise they could only be a result of parallel somatic mutation in those two individuals. An important consequence of this conclusion is that such sequences are faithful translations of the germline genes from which they derive; when they are compared to other related sequences, they can give us some idea of where in the descent the germline genes fall.

### The Meaning of Sequence Identities

In general, a somatic theory does not predict sequence identities, for few sequences are (in that theory) the products of unmutated germline genes. Recent models for the somatic generation of antibody diversity, however, have often supposed that many of the germline genes on which somatic mutation operates are themselves highly selected for particularly useful immunological functions, and thus are frequently expressed in an unchanged form. Clearly, we cannot take the mere existence of sequence identities as refutation of the somatic mutation theory.

Nevertheless, identical (that is, unmutated) sequences in the somatic mutation theory should bear a peculiar relation to the mutated sequences, which should stand out if enough related sequences are known. If it is supposed that the typical immunoglobulin chain expresses many somatic mutations, the identical sequences—which have not suffered these mutations—should fall unusually high on the evolutionary tree; that is, they should have undergone an unusually small number of mutations since their descent from their common ancestor with other closely related sequences. In contrast to these expectations of the somatic mutation hypothesis, the germline theory (which supposes that most sequences express unmutated germline genes) predicts that identical sequences should fall just like any other sequences in the evolutionary descent.

One case of apparently identical human $V_{\kappa I}$ sequences from different individuals (Dav and Fin) has been reported by Capra and Kunkel (1970). Both of these patients suffer from the rare disease hypergammaglobulinemia purpura, and their abnormal immunoglobulins share the same peculiar antiglobulin reactivity. The sequences of Dav and Fin are identical for 40 positions (see Table 2-2A), which include the first hypervariable region where all other sequences are very different from one another. Peptide mapping suggests identity of the remainders of the chains.

Dav and Fin differ in at least six nucleotide positions (13.1, 14.1, 14.2, 24.1, 34.2, 39.1) from all other $V_{\kappa I}$ proteins and thus have diverged by at least this number of mutations from any reasonable common ancestor with these chains. Like any other pair of definitely related sequences, they allow us to infer one more germline gene, by the same principles used in the previous sections.

Far from exhibiting an unusually small number of mutations, as might be expected if most other proteins suffered a large number of somatic mutations, Dav and Fin seem to be unusually divergent from other $V_{\kappa I}$ sequences. They are different in 14 nucleotide positions in the first 40 codons from the common $V_{\kappa I}$ ancestor (reconstructed according to the descent in Fig. 6-8C). The other $V_{\kappa I}$ proteins Ou, Hau, Paul, Au, Roy, Eu, Ag, and HBJ4 are different at 1, 1, 4, 4, 5, 6, 7, and 11 nucleotides, respectively. This distribution suggests that somatic mutation contributes only a small number of mutations, if any, to these proteins, for the most highly mutated of them are Dav and Fin, which have suffered no somatic mutations at all.

Sequence identities are a fundamental expectation of the germline theory, for it supposes that most sequences are the direct expressions of a limited (though large) number of germline genes. Indeed, *lack* of sequence identities may be used in principle to rule out a germline theory; for if enough sequences are studied and few or no identities found, the minimum number of genes which the theory must accommodate might be pushed beyond the limits of reasonableness. Figure 1-4 shows the minimum total number of different sequences inferrable (at the 95 percent confidence level) when given numbers of sequences are compared and no identities found. According to the germline theory, this minimum total number of different sequences is also (at least approximately) the minimum total number of genes. As can be seen from the graph, however, the number of sequence comparisons which must be performed before the number of inferrable genes would occupy an unreasonable proportion of a single average human chromosome is very large indeed. Conversely, an *abundance* of sequence identities might in principle rule out somatic mutation if, as more and more sequences are studied, a high proportion of the sequences are observed more than once. The fact that we have had to study so many proteins in order to find a single pair of putative identities among human L chains, however, makes it unlikely that we will soon have a complete catalogue of their germline genes, even if the germline theory is correct.

Such an exhaustive search for identities might be more fruitful in mouse L chains. Two probable pairs of identical mouse κ chains have already been reported (Hood et al., 1972). The mouse λ chains are an even more favorable case. No less than six identical mouse λ sequences have been reported by Weigert and his colleagues (1970), and all but one of the other five sequences (MOPC 315) differ from these by only a few mutations (see Table 2-3A). The low representation of these chains in normal mouse immunoglobulins fits this finding, for there might be few immunological tasks for which chains of such limited diversity are suited.

There is some pattern to their diversity: in three of the eleven proteins (including the unusually divergent MOPC 315) Asn replaces Ser at position 26. This suggests two nodes in their descent, thus two germline genes. However, this mutation might be a particularly favorable one repeatedly selected in different individual mice; hence no more than one germline gene can be inferred with any confidence.

These data fit the expectations of the somatic mutation theory, with identities and variants all diverging (more or less) from a single genealogical node representing a single germline gene. Since there has been little mutation of any sort in these proteins, this putative somatic mutation must be very limited. Although this limitation might be attributed to the short lifetime of the mouse, as suggested by Cohn (1970), the case of Dav and Fin above suggests (but does not prove) that the contribution—if any—of somatic mutation to the variation of human $V_{\kappa I}$ sequences is similarly small.

The germline theory's explanation of these results is that a limited number of mutations have accumulated during several duplications of the mouse $V_\lambda$ genes. The rate of mutation must be low relative to the rate of gene duplication to explain why so few of the mutations have been observed in different proteins. And the total number of genes must be large enough to explain why none of the sequence variants has been observed twice. Both of these suppositions must be regarded as ad hoc to some degree.

Studies of several additional variant mouse λ sequences might allow a satisfactory choice between these two explanations. If, on the one hand, repeats of several of the variant sequences are found, the case for somatic mutation in these proteins disappears. If, on the contrary, none of the new variants is a repeat of other variants (or at least shares some mutations with them), the number of genes that must be postulated for the germline explanation may become so large that it must be discarded as an unnecessarily complicated explanation of the facts. This limited amount of diversity then would have to be attributed to somatic mutation.

**Conclusions**

All three mechanisms of producing variation that I have considered—germline mutation, somatic mutation, and recombination—presumably contribute in some degree to antibody diversity. The theories differ over their relative importance. The last six years' work on immunoglobulin variability has shown that most of the variation observed between different immunoglobulin chains of the same family has occurred by germline mutation.

Almost no recombination has been found, either between different V-region subgroups, nor within the one subgroup ($V_{\kappa I}$) where it could have been detected. The somatic recombination theory, which (as a multigene theory) is subject to the same objections as the germline theory (see next chapter) without any evidence in its favor, thus no longer seems to be a viable alternative.

A choice between the germline and somatic mutation theories, however, is more problematical. The highly ordered genealogical relationships inferred in this chapter imply a relatively small contribution of somatic divergence and a correspondingly large number of germline genes. An even smaller contribution of somatic mutation is suggested by the nearly invariant mouse λ chains and by the relation of the two (probably) identical human proteins Dav and Fin to other human $V_{\kappa I}$ sequences. Of the many mutations that actually expressed immunoglobulin genes have accumulated, only the last few (if any) seem attributable to somatic mutation. But these few mutations may make the difference between an unworkable immune apparatus restricted to a limited, though large, number of germline genes, and a workable one able to call upon minor variants of these genes to meet an unexpected antigenic contingency. If so, somatic mutation may be said to be the essential component of antibody diversity, even if it is quantitatively minor.

A final choice between these two possibilities rests on further experiments. I have suggested that further studies on mouse L chains—particularly λ chains—might allow us to confirm or deny the residual contribution of somatic mutation which the present sequence evidence permits. Analogous studies with other chains or species might also lead to the same results.

**DNA-RNA Hybridization**

A completely independent way of checking and extending the conclusions of the genealogical evidence might be provided by DNA-RNA hybridization. Greenberg and Uhr (1967a and b) reasoned that if the somatic mutation theory is correct, the V genes present in different plasma-cell tumors will be different, and suggested that differences in the binding of immunoglobulin mRNAs to heterologous and homologous DNAs (that is, from different or the same tumors) might therefore support such a somatic theory. Cohen and Raska (1968; Cohen, 1967) demonstrated "antigen-specific" species of RNA in the spleens of mice injected with different red-blood-cell antigens by hybridization experiments with normal mouse thymus DNA; their suggestion that such experiments might support a multiple-gene hypothesis cannot be accepted, since different species of RNA would be expected in *any* theory of antibody diversity: it is in DNA differences that the theories are distinct.

Far more fruitful than these approaches, which focus on differences among RNA's and DNA's in different immune cells, would be experiments which depend on the similarities of many immunoglobulins to one another. It is likely that a V-region mRNA will be sufficiently homologous to a large proportion of the V genes in the same family to bind to them. The amount of RNA bound at saturation by a given amount of, say, sperm DNA would be a measure of the number of such homologous genes per haploid genome, and quantitative expectations of the germline and somatic mutation theories might in this way be checked. Particularly useful would be a measure of the number of mouse $V_\lambda$ genes, for in this case the germline theory predicts many times the single gene expected by the somatic hypothesis, and the sequences are so similar to one another that it is almost certain that they will cross-hybridize. Purified immunoglobulin mRNA might also be used to purify and visualize a complex immuno-

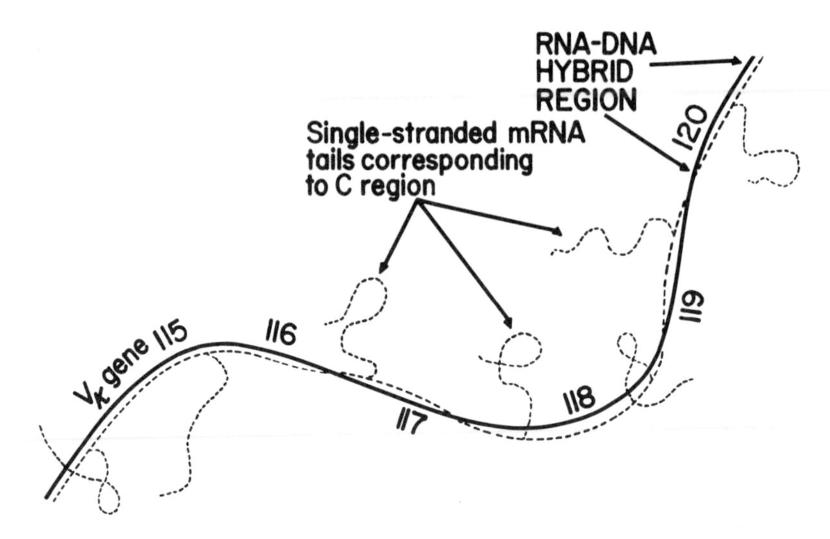


**Figure 6–16.** A hypothetical structure which might be visualized in the electron microscope by isolating hybrids between purified $\kappa$ messenger RNA and single-stranded DNA. Such a structure would arise only if the multiple $V_{\kappa}$ genes were in close tandem array; this structure would not be expected from the DNA network hypothesis (Chapter 9).

globulin locus, as shown in Fig. 6–16. Conclusive results from this type of experiment depend on the purification of immunoglobulin mRNA from the myriad other molecular species which are presumably present in the antibody-producing cell, a difficult technical feat.

# Chapter 7

## Species Specificity and Allotypes in V Regions: Evidence against the Germline Theory

I doubt that I am the only theoretical immunologist who has had occasion to regret the existence of the rabbit. Since the beginning of 1965, I had been predisposed toward the germline theory of antibody diversity and on the strength of that conviction had enthusiastically supported the heretical proposal of Dreyer and Bennett (1965) that the genes for V and C regions are separate in the germline. At the very time when this view was being vindicated on the basis of myeloma protein sequences (Dreyer, Gray, and Hood, 1967; Hood et al., 1967), other workers studying the structure of normal rabbit immunoglobulins were marshaling what seemed at the time an even stronger case against it, thus creating a paradox. These studies, which have been fully supported by further work, showed (*a*) that there are species-specific differences between the V regions in man and rabbit (Doolittle and Astrin, 1967), and (*b*) that there are allelic (allotypic) differences determined by rabbit $V_H$ regions (Koshland, 1967). It was reasoned that if the immunocompetent species harbor a large number of V regions, mutations would accumulate independently in each during phylogenetic and intraspecies evolution and hence could not be expected to be detectable.

This chapter will outline the arguments by which proponents of the germline theory seek to circumvent this awkward evidence. Only in a historical sense can the explanation advanced for species-specific differences be considered ad hoc: it is so reasonable that these differences can no longer be considered contrary to the germline theory. The allelic differences in rabbit $V_H$ regions, on the other hand, remain an important barrier to unreserved acceptance of the germline hypothesis.

### Species Specificity in V Regions

In Figs. 6-7, 6-9B, 6-11, 6-12B, 6-12D, and 6-13, the genealogical relation of V regions from various non human species to those from humans were displayed. Although many of these relationships may not be correct in detail, they demonstrate two facts well outside the limits of uncertainty of the reconstruction method. First, nonhuman sequences do not mix closely with human ones; rather, sequences from different species tend to form distinct, species-specific subgroups. For instance, there are no mouse sequences which fall within any of the three human $V_{\kappa}$ subgroups (see Fig. 6-7). From this we may conclude that many of the divergences between V regions, such as those within the $V_{\kappa}$ subgroups, have occurred subsequently to the early mammalian phylogenetic divergences, such as that between mouse and man. If, on the contrary, we were to suppose that all the V-region divergences preceded the phylogenetic ones, we would have to invoke some very special assumptions to explain why the surviving sequences chosen for study happen to assort into species-specific subgroups.

Secondly, the genealogical subgroups in different species do not correspond to one another on a one-to-one basis. From this we may conclude that V genes have undergone separate deletions and duplications in different species. For example, there is firm evidence in Fig. 6–7 for at least seven $V_{\kappa}$ subgroups (thus germline genes) in the mouse; since none of them falls within the human subgroups, and since there are only three of the latter, clearly there are at least four mouse V genes for which there are no human counterparts (or at least, whose human counterparts are almost never expressed).

When species specificity was first observed, it was taken as contrary to the germline view. The assumption implicit in this reasoning was that since the germline theory attributes diversity to gene duplication, and since diversity was presumably present in the common ancestor of all immunocompetent species, all V-region divergences should result from gene duplications *preceding* the phylogenetic divergences of the immunocompetent species. This assumption is simply not warranted: the generation of diversity by gene duplication can easily be envisioned as an ongoing process.

Species specificity would be a natural consequence of ongoing gene duplication and deletion. During the deletions and compensating duplications that intervene in some particular species line between the present and some ancient starting time, the deletions will eliminate by chance, one by one, all descendants of almost all the original germline genes, until eventually all the genes derive from a few of the original ones. The same would apply to another species line descending from the same original ancestor, except that the few original genes from which descendants survive in the one line would not be the same as those in the other. Species-specific subgroups would therefore ensue, as shown schematically in Fig. 7–1.

In accord with this view is the fact that species specificity is exhibited in

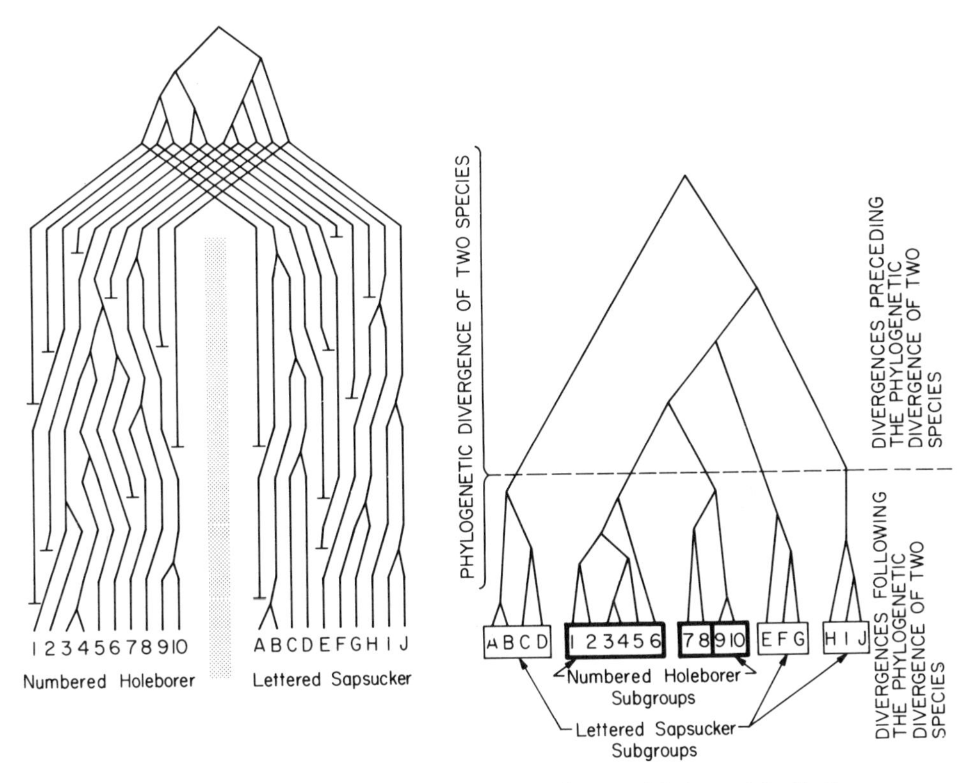


**Figure 7-1.** Evolution of species-specific differences by gene deletion and duplication. Taken from Smith, Hood, and Fitch (1971).

known multigene systems. For example, it is clear that ribosomal RNA genes are multiple and differ in their "spacer" portions from species to species (Appendix C). Also, it is clear from the above-mentioned lack of correspondence of V-region subgroups in different species that at least a limited amount of gene duplication and deletion must be admitted for antibody genes in any case.

## Rabbit $V_H$ Allotypes

At the outset, I shall admit that I use a double standard in interpreting the immunoglobulin allelic differences (or allotypes). For the vast majority, which must or can be assigned to C genes, I accept without elaborate critique the superficially obvious interpretation (Appendix B). They are accepted as amino acid replacements in the corresponding structural genes, and only one copy of each such structural gene is considered to occur per haploid genome. When, however, the obvious interpretation of an allotype conflicts with other theoretical considerations, I take pains to point out various alternative interpretations which are more congenial to my predilections. The rationale for this practice is that the assumptions that I accept uncritically in case of "normal" allotypes are either mutually confirmatory or else have not been used to develop any further arguments.

> Thus, the hypothesis that allotypes reflect allelic differences in structural genes allows a highly consistent picture of the genetic organization of these genes: they form three linkage groups corresponding exactly to the three gene families inferred from genealogical considerations in Chapter 9. And the supposition that these structural genes occur in only one copy per haploid genome has not been used at any point in this book to develop further arguments; in particular, this assertion is in no way necessary to show that V and C genes are not associated in the germline on a one-to-one basis (see Chapter 9).

The rabbit *a*1, *a*2, and *a*3 allotypes, more than any other body of experimental evidence, conflict with the germline theory of antibody diversity. In the remainder of this chapter I will subject these allotypes to a searching review and seriously suggest some alternative interpretations which, in any other case, I would probably dismiss as hopelessly ad hoc.

The *a*1, *a*2, and *a*3 allotypes are immunoglobulin determinants which are "isoantigenic": that is, they are detected by antisera raised in some rabbits by immunization with immunoglobulin from other rabbits. By a very sensitive test they are mutually exclusive; Oudin (1960b) tested 155 rabbits from an interbreeding population in which all three determinants occurred with appreciable frequency and found no rabbit with all three of them. If during the entire ancestry of these rabbits the accumulated incidence of chromosomes coding for more than one of these specificities were greater than about 3 percent, he would have been expected by chance to find such "triple phenotypes" in his sample.

The simplest interpretation of these data is clearly that *a*1, *a*2, and *a*3 are allelic forms of a single gene, for if there were multiple genes for each allotype, crossing over could give rise to chromosomes coding for more than one of them. But it is important to point out how big a locus might be expected to behave like a single gene. In the mouse the summed length of the known genetic map corresponds to an average recombination rate of only $2 \times 10^{-4}$ between markers separated by $3 \times 10^{4}$ base pairs—the length of about one hundred tandemly linked V genes (Sager and Ryan,

1961). Even this many genes, therefore, might have been scrambled too infrequently to fail the triple phenotype test. Moreover, we have no idea whether this recombination rate, computed for relatively distant genes, can be meaningfully interpolated to the miniscule chromosomal segment occupied by even hundreds of immunoglobulin genes: the actual rate may be orders of magnitude lower (or, of course, higher).

A number of independent lines of evidence lead to the conclusion that *a*1, *a*2, and *a*3 specificities are determined by the $V_H$ region. They are found on the H chain of all four known classes of immunoglobulin in the rabbit (IgG, IgM, IgA, and IgE); the allotypic determinants in different classes are the same or at least cross-reactive (Feinstein, 1963; Kindt, Steward, and Todd, 1969; Kindt and Todd, 1969; Lichter, 1967; Lichter et al., 1970; Pernis et al., 1969; Segre, Segre, and Inman, 1969; Sell, 1967; Stemke and Fischer, 1965; Todd, 1963; Todd and Inman, 1967; Utsumi, 1969). From the considerations in Chapter 9, they must be assigned to $V_H$ regions, because these are the only structures shared by $\gamma$, $\mu$, $\alpha$, and $\epsilon$ chains. In IgG (and presumably in other classes as well) they are located in the Fd portion ($V_H + C_H1$) of the H chain (Feinstein, Gell, and Kelus, 1963; Micheli, Mage, and Reisfeld, 1968; Stemke, 1964).

There are reproducible, mutually consistent amino acid compositional differences between $\gamma$, $\mu$, and $\alpha$ chains of different allotypes (Inman and Reisfeld, 1968; Koshland, Davis, and Fugita, 1969; Koshland, Reisfeld, and Dray, 1968; Prahl and Porter, 1968; Wilkinson, 1969b), and these differences are consistent with, and largely account for, compositional (Koshland, 1967; Prahl and Parker, 1968) and sequence (Wilkinson, 1969a) differences within the first 34 residues of rabbit $\gamma$ chains. The major average sequences found by Wilkinson for these positions of *a*1 and *a*3 $\gamma$ chains can be compared in Table 2-4A. There are, by contrast, no apparent differences between *a*1 and *a*3 $\gamma$ chains in their $C_H1$ regions (Wilkinson, 1970). In addition, similar N-terminal oligopeptide differences have been described for $\alpha$ and $\gamma$ chains (Wilkinson, 1969a and b).

Finally, in double heterozygotes for these markers and various pairs of markers determined by the $C_\gamma$ gene, individual IgG molecules express only or predominantly *parental* combinations of markers (Chapter 9), a highly fortuitous circumstance if *a*1, *a*2, and *a*3 were determined by some secondary feature of these molecules, such as their carbohydrate moieties. Let us then accept in what follows that *a*1, *a*2, and *a*3 determinants reflect differences in the corresponding $V_H$ regions.

It is clear, despite Oudin's failure to detect triple phenotypes, that there are at least several different germline $V_H$ genes in each of the three possible homozygotes, and thus that each *a* "locus" is in fact a family of several, presumably closely linked, $V_H$ genes. By Ouchterlony analysis, Oudin (1966) found, for instance, that each of the antiallotype antisera detects several specificities located on separate immunoglobulin molecules. Moreover, at least two or three major average sequences can be detected in each of the three homozygotes, as shown in Table 7–1. Finally, every normal rabbit contains a small proportion of *a*-negative molecules which fail to react with any antiallotype antiserum. (Strictly speaking, however, these need not be attributed to different germline genes, since it is possible that they represent somatic variants in which the *a* determinants have been altered.)

The interpretation of these facts by the somatic mutation theory is not entirely straightforward. It can be assumed that there are a few linked $V_H$ genes in the rabbit genome, and that there are three major types of chromosomes, in each of which most of these genes have undergone germline mutations. These genes can reasonably be supposed to be so tightly linked that the linked differences on one chromosome behave as if they were

**Table 7–1.** Evidence for multiple subgroups (germline genes) in rabbits homozygous for $V_H$ allotypes (Wilkinson, 1969a, 1970).

A. Yield of Short N-terminal pronase peptides from $\gamma$ and $\alpha$ chains of various allotypes (moles per mole of H chain).

| | $\gamma$ chains of allotype– | | | $\alpha$ chains of allotype– | |
|---|---|---|---|---|---|
| Peptide | *a*1 | *a*2 | *a*3 | *a*1 | *a*2 |
| *PCA-Ser-Val-Glu | 0.56 | 0.30 | – | 0.55 | – |
| *PCA-Ser-Leu-Glu | 0.14 | 0.30 | 0.50 | 0.14 | 0.43 |
| *PCA-Glu-Gln | – | 0.07 | 0.26 | – | 0.22 |
| *PCA-Ser-Ser | – | – | – | 0.14 | 0.16 |

*PCA is pyrrolidone carboxylic acid, a cyclized form of Glu or Gln

B. Evidence for two subgroups of $V_H$ regions from *a*3 homozygous rabbits. Three amino acid replacements and one deletion (boxed positions) distinguish these two variants; these linked differences are most unlikely to arise by somatic mutation in many independent cell lines, and constitute evidence for at least two germline $V_H$ genes in *a*3 homozygotes.

MAJOR SEQUENCE: *Z [– S] L E E S G G D L V [K] P G [A] S L T L T C

MINOR SEQUENCE: *Z [E Q] L (E,E,S,G,G,$^{V}_{D}$,L/V,[K̃,]P,G,[Ã,]S,L)T L T C

*Z is pyrrolidone carboxylic acid.

˜Residue was *not* present, but its replacement could not be determined.

determined by a single gene. This interpretation is diagrammed in Fig. 7–2A. It is somewhat puzzling, under this hypothesis, however, that so many linked differences are assorted into three distinct chromosomes. Multiple linked differences, such as those observed between the average sequences of *a*1 and *a*3 $V_H$ regions, are usually thought to arise in separate noninterbreeding populations, as depicted in Fig. 7–2B. The hypothesis that differences between *a*1, *a*2, and *a*3 arose in isolated populations allows an alternative explanation based on gene duplication and deletion, as depicted in Fig. 7–2C. Since gene duplication and deletion require multiple crossover events, one would expect that chromosomal differences would be scrambled as fast as they arose in an interbreeding population; hence this hypothesis, like that of Fig. 7–2B, requires the supposition of isolated breeding populations.

In the rabbit $C_\gamma$ gene, which is linked to the *a*1, *a*2, and *a*3 alleles, only two allelic amino acid replacements have so far been detected (see Appendix Table B–7). The hypothesis in Fig. 7–2B, then, seems to require that few differences accumulated in the $C_\gamma$ genes while many accumulated in the linked $V_H$ genes, even though C genes in general seem to accumulate mutations as rapidly as V genes. This apparent paradox does not occur in the gene duplication and deletion model (Fig. 7–2C), for it can reasonably be supposed that radical shifts in chromosomal populations of $V_H$ genes can occur in the time required to accumulate only a few mutations.

The interpretation of these allotypes by the germline hypothesis is complicated chiefly by the large number of genes involved, which I shall assume to be no smaller than one hundred. With so many genes, the hypothesis that detectable allelic differences arose in three distinct chromosomes in an interbreeding population (Fig. 7–2A), which is already doubtful in the case of a few genes, becomes untenable. It is also highly unlikely that they could arise in one hundred genes by the simple accumulation of mutations in isolated populations (Fig. 7–2B), for these allelic differences would accumulate independently in each of the hundred genes, and the resulting multiplicity of independent antigens recognized by antiallotype antisera would undoubtedly preclude their detection. The gene duplication and deletion model (Fig. 7–2C), however, represents a reasonable explanation for these allelic differences; for once the hypothesis that they accumulated in isolated populations is put forth, they become no different in principle from the species-specific differences discussed earlier.

There are a few alternative explanations of the rabbit $V_H$ allotypes which

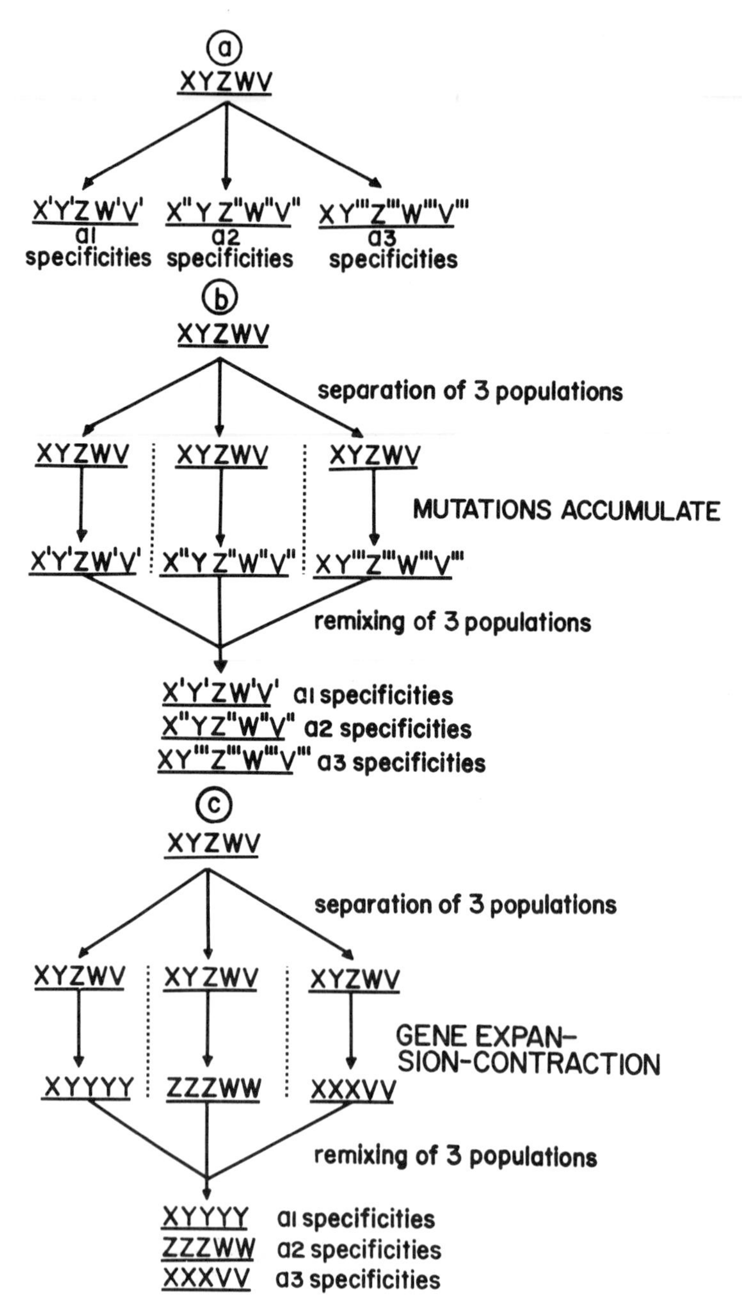


**Figure 7-2.** Three hypotheses for the evolution of rabbit $V_H$ allotypes: (*a*) simple mutation in one interbreeding population; (*b*) mutation in separate populations; (*c*) gene duplication and deletion (expansion-contraction) in separate populations.

will be mentioned only briefly. (1) It is possible that they arose, even in an interbreeding population, by democratic gene conversion, which was described in Chapter 5. (2) It is possible that they arose, again in an interbreeding population, by *intrachromosomal* gene duplication and deletion: that is, the chromosomal isolation attributed above to geographic isolation of subpopulations of rabbits might be attributed instead to some prohibition on interchromosomal crossover between $V_H$ genes. This might have occurred, for example, if small inversions or rearrangements made the $V_H$ genes on the three different chromosomes unable to pair, or if some specific block to effective synapse in this region is present. (3) The allelic differences might reside, not in the $V_H$ genes, but in the gene(s) for the "joining mechanism" (see Chapter 9) for linking $V_H$ and $C_H$ genes; different alleles might prefer to link different populations of $V_H$ genes.

The foregoing ad hoc reconciliation of rabbit $V_H$ allotypes with the germline theory does not seem sufficiently awkward to be dismissed. For even the somatic mutation theory has some difficulty explaining them, and what seems to me a plausible way out of the relatively minor difficulties they pose for the somatic theories—namely, the supposition that they arose in isolated populations of rabbits—is equally a way out of the major difficulties they pose for the germline theory.

In Chapter 9 we shall look at Smithies' hypothesis of branched DNA networks for the immunoglobulin loci. This theory permits two new possible explanations for the rabbit allotypes: (1) they might be coded by regions of low lateral multiplicity in which allelic differences have accumulated; or (2) they might, even without hypothesizing isolated noninterbreeding populations of rabbits (or nonsynapsing chromosomes), arise by gene duplication and deletion, which in this model would occur exclusively intrachromosomally.

A great deal more work is required before these markers can be fully interpreted. As as start, homogeneous rabbit antibodies might be used to sort into families the multiple determinants evidently being detected by antiallotype sera, in much the same way that homogeneous human and mouse myeloma proteins have permitted the assignment of the *Gm* and *Ig* determinants to different closely linked genes. The amino acid sequence concomitants of these determinants might then be sought in more detail than has hitherto been possible. Lastly, the distribution of these resolved determinants might be determined in closely related species, such as hares and pikas, in order to elucidate the evolutionary origin of these allelic differences.

An allelic V-region difference with very different characteristics has been reported by Edelman and Gottlieb (1970) for mouse $\kappa$ chains. In autoradiograms of partially resolved cysteine-containing tryptic peptides of normal $\kappa$ chains from several inbred strains, some strains are found to have a spot not found in others. It is not fully resolved, but is distinguishable, from another spot occurring in all strains. In mating studies the presence of this spot behaves as a simple mendelian trait which is dominant over its absence. In contrast to the rabbit allotypes just discussed, this allotype affects only a small proportion of $V_\kappa$ regions. The authors imply that the peptide responsible for this spot occurs in more than 10 percent of the $\kappa$ chains, based on its yield of labeled cysteine relative to that from a $C_\kappa$-region peptide. However, because the spot is poorly resolved from another occurring in all mice, this yield can only be taken as a preliminary estimate of the actual frequency of chains carrying this allotype. The presence or absence of this spot in various inbred strains does not not correlate with their $C_H$-region allotypes (Herzenberg, McDevitt, and Herzenberg, 1968). I do not think this allotype can be taken as evidence contrary to the germline view. For the assumption of Edelman and Gottlieb that this theory requires at least one thousand mouse $V_\kappa$ genes is unwarranted. Presumed identities have already been found among these chains, indicating that the total number of variants may be small. Although its value as evidence against the germline theory is dubious, this marker–if it is confirmed by further study–might well prove to be a valuable approach to the genetics of mouse immunoglobulins.

Gibson, Levanon, and Smithies (1971) have carried out an extensive search for such V-region allotypes in normal human $\kappa$ and $\lambda$ chains, by looking for differences between L-chain tryptic peptides obtained from several individuals. These peptides showed remarkable similarities (both quantitative and qualitative) from person to person, and no V-region differences were detected.

Both germline and somatic mutation theories are capable of explaining the V-region allotypes with the aid of some special, though plausible, assumptions. But where the somatic theory requires relatively limited special assumptions, the germline theory, by virtue of the larger number of germline V genes it supposes, invokes a more extensive concatenation of circumstances. For example, in the case of rabbit $V_H$ genes, radical shifts of large chromosomal populations of genes must be assumed for the germline theory, whereas in the somatic view only a limited amount of gene deletion and duplication is required to bring about the observed allelic differences. In this sense, these allotypes weigh against the germline theory.

Table 7–2 summarizes the various lines of evidence brought to bear on the origin of antibody diversity in this and the previous chapters. Contrived explanations are outlined for each fact which seems awkward for one of the theories, and those which seem particularly cumbersome are boxed for emphasis. The facts appear to me to render somatic recombination a most unlikely explanation of antibody diversity. The choice between somatic and germline mutation is much more difficult to make, for there is some evidence for both views. It may seem wisest under these circumstances to embrace some compromise, in which most mutational differences in V regions accumulate presomatically and the last few somatically, as outlined at the end of Chapter 6. But I choose instead to advocate a "pure" germline theory, in which somatic mutation occurs rarely (with the same frequency in V genes as in C genes) and is not essential for the function of the immune system.

In the eight years since I first took an interest in antibody diversity, more and more direct evidence has accumulated in favor of the germline theory, until the identical V regions discussed in the last chapter have virtually established that that hypothesis' basic tenet is fullfilled in some cases. The case for somatic mutation essentially consists in facts which, through a series of successive inferences, make a pure germline theory seem implausible. Indeed, it was this sort of "fact" which originally sparked interest in the origin of antibody diversity: when Breinl and Haurowitz (1930) convinced immunologists that preformed antibodies to arbitrarily chosen, even artificial, antigens could not exist in an animal nonetheless capable of responding to them, they initiated over forty years of speculation on unconventional means by which antibody diversity might be generated. But neither this nor any other argument yet raised against a pure germline theory is direct and immediately compelling, and I suspect that immunologists will in the end settle for the conventional view that antibody diversity, like other examples of biological diversity, is carried in the germline.

**Table 7–2.** Compendium of theories of antibody diversity.

| Elements of theories | Germline theory | Somatic recombination theory | Somatic mutation plus selection theory | Somatic hypermutation theory |
|---|---|---|---|---|
| Origin of diversity | Faithful translation of duplicated, mutated germline V genes | Somatic recombination among duplicated, mutated germline V genes | Ordinary low-level somatic mutation plus selection for variants | Somatic hypermutation |
| Restriction of diversity to V regions | Only V genes are extensively duplicated | Only V genes are extensively duplicated and recombined | Selection for V-gene mutants, no selection for C-gene mutants | Hypermutating apparatus recognizes only V genes |
| Number of genes proposed for human $V_\kappa$ | $> 425$ | 10–100 freely recombining genes per subgroup | 1 per subgroup | 1 per subgroup |
| Percent of average human chromosome occupied by $V_\kappa$ genes | $> 0.08$ | 0.006–0.06 | 0.0006 | 0.0006 |

| Facts | Ad hoc explanations (significant ones are boxed) | | | |
|---|---|---|---|---|
| Highly ordered $V_{\kappa}I$ genealogy | | | Increase number of $V_{\kappa}I$ genes | Increase number of $V_{\kappa}I$ genes |
| No recombination *between* major subgroups | | [boxed] Divide V genes into subgroups between which no recombination occurs | | |
| No recombination *within* human $V_{\kappa}I$ subgroup | | [boxed] Divide $V_{\kappa}I$ into sub-subgroups between which no recombination occurs | | |
| No presomatic mutations evident in human $V_{\kappa}II$ and $V_{\kappa}III$ | Intermediate ancestors during duplication of these genes accumulated few mutations | | | |
| Rabbit $V_H$ allotypes | [boxed] Rapid gene deletion and duplication shifts $V_H$ genes in isolated subpopulations of rabbits | [boxed] Same as for germline theory | | |

(continued)

**Table 7–2 (continued)**

| Facts | Ad hoc explanations (significant ones are boxed) | | | |
|---|---|---|---|---|
| Elements of theories | Germline theory | Somatic recombination theory | Somatic mutation plus selection theory | Somatic hypermutation theory |
| Species specificity | Extensive gene deletion and duplication | Same as for germline theory | Limited gene deletion and duplication | Limited gene deletion and duplication |
| Mouse $V_\lambda$ identities | Duplications of $V_\lambda$ gene with few mutations | Same as for germline theory | Few somatic mutations, frequent expression of unmutated germline genes (because of short lifetime of mouse?) | Same as for somatic mutation plus selection |
| Identical human $V_\kappa I$ sequences Dav and Fin, which do not seem to express an unusually small number of mutations | | | Few somatic mutations, even in long-lived species; occasional expression of unmutated germline genes | Same as for somatic mutation plus selection |

| Future experiments | Predictions | | | |
|---|---|---|---|---|
| Very extensive human or mouse $V_\kappa$ data | Repeated identities | Exact recombinants | Only occasional identities | Only occasional identities |
| More mouse $V_\lambda$ data | Repeats of *all* variants | Exact recombinants | Repeats only of 6 present identities | Repeats only of 6 present identities |

# Chapter 8

## The Clonal Selection Theory

The importance of antibody diversity as an experimental problem in recent years has led to an exaggeration of its importance as an evolutionary problem. I suggest that the crucial step in the evolution of adaptive immunity was the emergence of the mechanism which allowed cells to become clonally specialized for the expression of a single antibody specificity. Clonal specialization seems to be the only plausible way in which preadaptive immunoglobulin variants can be adaptively expressed and was therefore probably necessary before antibody diversity could confer real selective advantage on the organism. Under these conditions, an ever-present tendency toward diversification (which in most systems is suppressed because diversity does not by itself carry any definite survival value—and may, indeed, be deterimental in that it dilutes functional with nonfunctional variants) was, in the case of immunity, released and encouraged by natural selection, so that the high level of antibody diversity observed today was reached very rapidly.

The presumed minimal elements of a theory of adaptive or selective (that is, antigen-specific) antibody synthesis are shown in Fig. 8-1. The same elements are applicable whether the synthesis of antibody is to be selectively enhanced (as in the antibody response to an exogenous antigen) or depressed (as in tolerance to endogenous antigens). Antigen is recognized by a receptor, which must somehow be linked to the synthesis of a variant of immunoglobulin whose antigen combining site is specific for the same antigen.

The clonal selection theory of Burnet (1959) incorporates all these elements in an entirely natural, uncontrived way, as outlined in Chapter 5. In only one major particular must Burnet's original theory be changed to fit the facts. He envisioned a "somatic mutation" event as committing a cell, and its progeny, to the expression of a single antibody specificity. It is clear now, as will be discussed in the next chapter, that the commitment event is not a simple mutation, but rather a complex happening in which the genes to be expressed are selected from among an extensive array.

The three fundamental postulates of the clonal hypothesis are as follows: (1) the antigen receptor site is identical to the antigen combining site; (2)

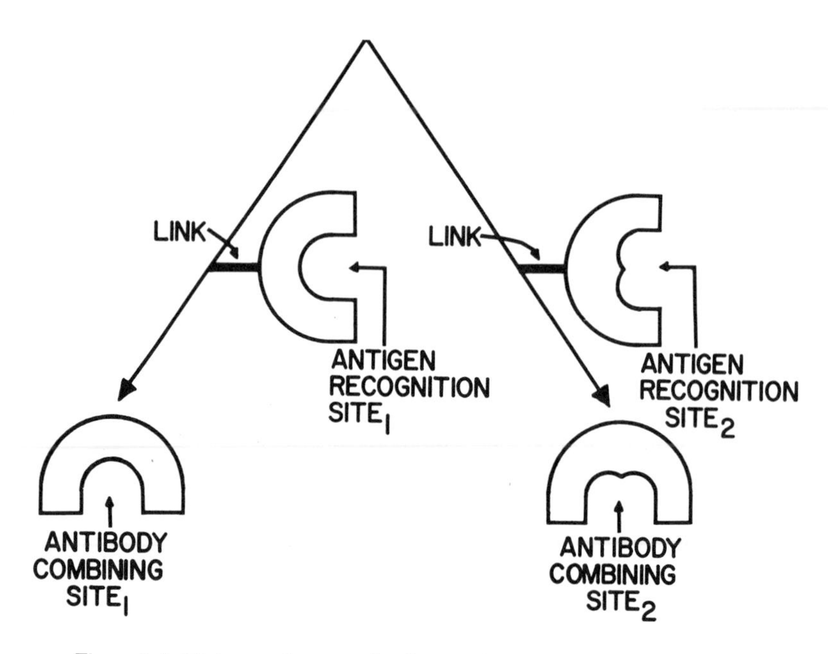


**Figure 8–1.** Minimum elements of a theory of selective antibody synthesis.

the specific link between the antigen receptor site and the synthesis of the corresponding immunoglobulin is mere physical coexistence in the same cell; individual cells, that is, are specialized for the synthesis of a single species of antibody, which serves as both antigen receptor site and antigen combining site; and (3) the cell specialization postulated in (2) is inherited and therefore clonal. I shall argue that the alternatives to these postulates are evolutionarily unlikely on a priori grounds, and will also discuss some of the experimental evidence pertinent to Burnet's theory.

*Postulate* (*1*): Each antigen receptor site is identical to the antibody combining site whose synthesis it controls. The combining specificity of the antigen receptor site must be extraordinarily similar to that of the antigen combining site to explain the fact that all the immunoglobulin synthesized in response to a given antigen has a high affinity for that antigen. In other words, the range of antigens with which a given antigen receptor site reacts effectively must be almost identical to the range of antigens with which the corresponding antibody reacts. Makela (1970; see also Francis and Paul, 1970; Mitchison, 1967) has recently reviewed the evidence for a close parallelism between the inferred properties of the antigen recognition site and those of the antigen combining site of the antibody produced. No such identity of reactivities is observed when two different antibodies directed against the same antigen are compared; for example, Jaton and his colleagues (1970) have shown that some individual rabbit antibodies directed against type VIII pneumococcal polysaccharide cross-react strongly with type III polysaccharide, while others fail to cross-react at all. It is unreasonable to suppose that the antigen receptor and antigen combining sites evolved independently in such an exactly parallel fashion, especially with respect to their reactivities to artificial antigens which could not have played a role in their evolution. Rather, it seems quite safe to conclude that the two sites are identical, and that the molecular species which recognizes antigen is in fact the corresponding antibody.

*Postulate* (*2*): Cells are specialized. If, contrary to the clonal selection theory, individual cells are not specialized for the production of a single molecular species of antibody (and thus antigen receptor), there must be some subcellular mechanism by which antigen specifically turns on the appropriate antibody structural genes, and not the other genes present. As diagrammed schematically in Fig. 8-2, subcellular selection would involve at least one other recognition system (and two complementary sites) for each antibody. It is difficult to imagine the necessary coordinate evolution of these sites if this subcellular selection is to be specific for the antigen; and it can be concluded, I think, that the link between the antigen receptor

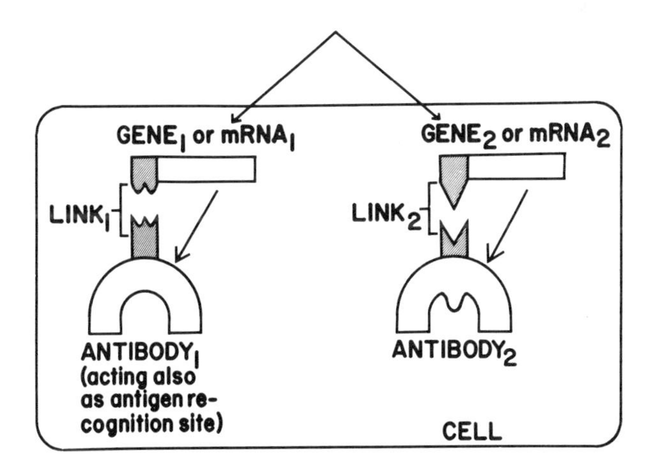


**Figure 8-2.** A minimum model for subcellular selection by antigen. This scheme is unlikely, because it seems to require additional complementary recognition sites for each antibody the cell can make. It is much simpler to assume that the specific link is mere physical association in the same cell, brought about because different antibodies are sequestered in different cells.

and the structural genes it specifically controls is mere physical association, presumably because they are present in the same cell. A given cell, according to the clonal selection theory, can only synthesize one species of antibody, which acts in some way as an antigen-triggered control for its own synthesis. The elimination of antibodies directed against endogenous antigens is also easy to understand by supposing this cell specialization, which allows the forbidden antibodies to be purged by removing the cells committed to them.

Cell specialization, the key proposal of the clonal selection theory, is not entirely compelling on a priori grounds. Subcellular selection mechanisms *can* be visualized which do not require that the "link" involve an extra pair of specific complementary recognition sites; all that the above reasoning demands is that some physical link remain between the antibody as receptor and the particular genes responsible for its synthesis. While segregation of different antibodies into different cells (as proposed in the clonal selection theory) is certainly the easiest way to meet this requirement, other ways can be imagined. For instance, nascent antibody molecules might remain associated with messenger RNA, and RNA with DNA, in such a way that the reaction of the nascent antibody with antigen could derepress transcription and/or translation. But such mechanisms are exceedingly awkward in the framework of our understanding of protein synthesis.

*Postulate* (*3*): Cell specialization is inherited (clonal). Antigen might control the synthesis of antibody at several different levels: it might influence the replication of the structural genes encoding the corresponding antibody by controlling the proliferation of the cells in which those genes are active; alternatively or additionally, it might control the transcription or translation of those genes. There is strong evidence that at least some of the response to an antigen results from proliferation of the cells synthesizing the complementary antibodies. If antigen specifically controls the proliferation of antibody-forming cells, the cell specialization hypothesized in Postulate (2) must be heritable. This heritable specialization also suggests a simple explanation of the secondary response. During the cell proliferation of the primary response, some of the progeny—each harboring the same specificity which was originally responsible for the stimulation of its progenitor—wait in large numbers for a second contact with the same antigen.

A number of aspects of the immune response are easily understood in terms of the population dynamics of cells which are under the selective influence of antigenic stimuli. It is often observed that in the course of an immune response the affinity of antibody for the antigen increases. This can be explained as the increasingly selective stimulation of subpopulations of cells specialized for more avid antibody as the antigen becomes scarcer and scarcer and the duration of the selection continues. Similarly, it is observed that when an animal is primed with a certain antigen and boosted with a related (but not identical) antigen, the antibody produced has more affinity for the first (priming) antigen than that which is produced if the second antigen is used both for priming and for boosting. This influence of priming antigen on the response to a related antigen is called "original antigenic sin." It is explained if the second antigen is thought to restimulate a substantial subpopulation of the memory cells produced by the first antigen, which happen to cross-react with it. The population of molecular species of antibodies (as measured by "anti-idiotype" antibodies, presumed to react with V-region determinants) seems also to change in the course of the immune response, suggesting again a corresponding shift in the population of antibody-producing cells (Daugharty et al., 1969; Eichmann et al., 1970; Oudin and Michel, 1969; Hopper, MacDonald, and Nisonoff, 1970; MacDonald and Nisonoff, 1970). Although all these changes are easily explained by the clonal selection theory, alternative interpretations may be imagined. This subject has been reviewed by Siskind and Benacerraf (1969).

## Evidence Pertaining to the Clonal Selection Theory

> An idea is only an idea, while a fact is only a fact.
>
> —Jean Piaget*

The clonal selection theory immediately suggested the means by which it could be tested. The majority of the experiments designed to test the hypothesis have conformed to its expectations, but a number have proved awkward for the clonal view. Whether one chose to interpret the contrary results as telling counterinstances or puzzling anomalies that would eventually work themselves out depended very much on one's theoretical predilections. As will become evident shortly, my own are all in favor of the theory. Many of those who are most directly concerned with testing Burnet's hypothesis take the few contrary findings more seriously than I

*Quoted in H. Ginsburg and S. Opper, *Piaget's Theory of Intellectual Development* (Englewood Cliffs, N.J.: Prentice-Hall, 1969), p. 2.

do, however, and several of the clonal selection theory's basic tenets—including antigen-independent cell commitment and the "forbidden clone" theory of tolerance—are under attack.

As Burnet (1967) has pointed out, myelomatosis constitutes almost a categorical demonstration of Postulate (3), since plasma-cell tumors are stably committed to the production of a single molecular species of immunoglobulin; in the mouse, the stability of this commitment has been demonstrated after multiple transplant generations (Potter, Appella, and Geisser, 1965; Potter et al., 1964; Potter and Lieberman, 1967).

An extensive body of data on normal immunologically active cells supports, with some reservations, the hypothesis that individual cells are committed to one each of all the possible H and L chains which the organism is capable of synthesizing, both at and prior to the stage of active synthesis and secretion of antibody. Active antibody-producing cells generally synthesize only one H or L chain class or subclass, one antibody specificity (in animals responding to more than one antigen), and (in heterozygotes) only one of two allelic forms (Table 8-1).

The restriction of single cells to one of two allelic forms is particularly interesting; allelic exclusion is observed in sex-linked genes in females, but the generally observed case for autosomal genes is that both alleles are expressed in each cell (see review by Brown and Dawid, 1969). In animals heterozygous at both H- and L-chain loci, there are individual molecules with each of the four possible combinations of parental alleles (Dray, Young, and Nisonoff; 1963). But, as will be discussed in Chapter 9, in rabbits heterozygous at both $V_H$ and $C_\gamma$ loci, only parental combinations of alleles are expressed in individual $\gamma$ chains.

Scattered reports of exceptions to the one-cell/one-immunoglobulin rule are listed in the footnotes to Table 8-1. Many other cases of apparent double-producing cells have been reported and subsequently criticized, and it is possible that the occurrence of some of these seemingly rare cells may similarly arise from an error in interpretation. It should be noted that the clonal selection theory is not incompatible with an occasional double-producing cell, either because of a stable commitment to two molecular species, or a recommitment from one species to another (both being detectable during the switching). The case of established human lymphoid cell lines which produce more than one type of immunoglobulin (Table 8-1, note a) will be dealt with later in this chapter.

The specialization of immunocompetent cells to a single species of antibody during their final production stage, however, does not constitute a compelling confirmation of the clonal selection theory. For once the possibility of some subcellular control mechanism is admitted, unipotency of cells in this phase might follow from plausible assumptions. The essence of Burnet's hypothesis, its point of departure from other possibilities, is that cell specialization is *not antigen dependent.* In order to test it more directly, we must ask if immunocompetent cells are specialized for a single molecular species of antibody *before* they are stimulated by antigen. Although the question is simple to ask, experimentally it has proved exceedingly difficult to answer. Before the problem can be intelligently appreciated, it is necessary to discuss what has been discovered about the cells which respond to an antigenic stimulus and how their response is manifested.

It was shown by Claman, Chaperon, and Triplett (1966), and since confirmed in many different systems (Cudkowicz, Shearer, and Priore, 1969; Miller and Mitchell, 1968; Mitchell and Miller, 1968; Nossal et al., 1968; Schimpl and Wecker, 1970; Schlesinger, 1970; Shearer, Cudkowicz, and Priore, 1969), that a vigorous antibody response to at least some antigens requires the cooperation of two cell types, which have been called T cells (thymus dependent) and B cells (thymus independent; found abundantly in bone marrow; and in the chicken dependent on another lymphoid organ, the bursa of Fabricius). Both cells are at least to some degree specific for the antigen, in that both must be exposed to it (Shearer and Cudkowicz, 1969) and neither is effective if the donor animal is tolerant to it (Chiller, Habicht, and Weigle, 1970). It is the B cells alone, however, which eventually produce antibody (Jacobson, L'age-Stehr, and Herzenberg, 1970; Miller and Mitchell, 1968; Mitchell and Miller, 1968; Nossal et al., 1968; Schimpl and Wecker, 1970; Schlesinger, 1970). In antihapten antibody responses, the T cells may be directed solely to the carrier moiety of the hapten-carrier conjugate (Katz et al., 1970a and b, 1971; Mitchison, 1969; Raff, 1970; Rajewsky et al., 1969).

Apart from their role in the humoral antibody response, T cells are thought to be the effectors of cellular immunity. This type of immunity differs from humoral immunity in that it cannot be transferred to another animal by serum, but only by living lymphoid cells. Like humoral immunity, cellular immunity is specific for the sensitizing antigen and exhibits specific tolerance and a specific accelerated, enhanced secondary response. The rejection of foreign organ transplants is thought to be effected by cellular immunity, as is the "delayed" hypersensitivity to some antigens

**Table 8–1.** Specialization of cells at the antibody-production stage.

| Study | Species | Specialization with respect to – | | |
|---|---|---|---|---|
| | | class | allotype | specificity |
| Bernier and Cebra (1965) | Man | $\kappa,\lambda;\gamma,\alpha$[a] | | |
| Cebra, Coldberg, and Dray (1966) | Rabbit | $\gamma,\alpha,\mu$ | $a1,a2$ ($V_H$) | |
| Nussenzweig et al. (1968) | Guinea pig | $\gamma_1,\gamma_2$[b] | | |
| Makela (1967) | Rabbit | | | Phages T2,T5,P22[d] |
| Pernis and Chiappino (1964) | Man | $\kappa,\lambda$[a] | | |
| Gershon et al. (1968) | Mouse | | | Rabbit, camel RBC<br>Poly-alanine, arsanilate[e] |
| Marchalonis and Nossal (1968) | Mouse | Electrophoretic mobility of presumed immunoglobulin from single cells | | |
| Luzzati, Tosi, and Carbonara (1970) | Mouse | Electrophoretic mobility of presumed immunoglobulin from a single node of lymphoid cells (presumed to be a single clone) in repopulated spleen in irradiated animal | | |
| Pernis et al. (1965) | Rabbit | | $a1,a2,a3$($V_H$)<br>$b4,b5,b6$($C_\kappa$) | |
| Green, Vassalli, and Benacerraf (1967a) | Guinea pig, rabbit | | | DNP,BSA,BGG[e] |
| Green et al. (1967b) | Rabbit | | | DNP-polylysine, albumin[e] |
| Weiler (1965) | Mouse | | $Ig$-$1^b$,$Ig$-$1^a$($C_{\gamma 2a}$) | |
| Bernier et al. (1967) | Man | $\gamma_1,\gamma_3$[a] | | |
| Nossal and Lederberg (1958) | Rat | | | Two Salmonella flagellar antigens |
| Cosenza and Nordin (1970); Nordin, Cosenza, and Sell (1970) | Mouse | $\gamma,\mu$[c] | | |
| Guillien, Avrameas, and Burtin (1970) | Rabbit | | | Two determinants on same molecule[e] |
| Merchant and Brahmi (1970) | Rabbit | $\gamma,\mu$ | | |

*Contrary reports:*

[a]Human lymphoid cell lines established in cell culture often produce low levels of immunoglobulin. Some lines show the same specialization observed in single antibody-producing cells (Matsuoka et al., 1968, 1969; Takahashi et al., 1969b), but others are observed to produce both $\gamma$ and $\alpha$ chains (Takahashi et al., 1968, 1969a) or $\gamma$ and $\mu$ chains (Feingold, Fahey, and Dutcher, 1968).

[b]Bozzi et al. (1969) directly contradict this result of Nussenzweig et al.

[c]In these experiments, a low level (about 1 percent) of cells produced both $\gamma$ and $\mu$ chains during the period in which IgM antibody was declining in the serum and being replaced with IgG antibody. The IgM to IgG switch is discussed in the text.

[d]This is a repeat, with very different results, of experiments performed by Attardi et al., (1959, 1964), in which a high level of double-producing cells was observed. Makela found not a single such double-producer.

[e]In these experiments, the two immunizing antigens were introduced on the same immunogenic molecules or particles in order to increase the likelihood of double-producers. In a similar experiment with different results, Michael and Marcus (1968) found that when two antigens were included in the same complex, double-producing cells were frequent; in control experiments in which the antigens were on separate particles, no double-producing cells were observed.

(for example, tuberculin and its derivatives) following certain procedures of immunization.

The cell dynamics of the immune response will concern us only insofar as it bears on the question of cell specialization proposed in the clonal selection theory. In assessing whether this specialization of cells precedes antigen stimulation, it is important to differentiate between B and T antigen-reactive cells. It is also crucial to the argument to differentiate between unprimed and primed cells, because specialization in the latter could depend on the original antigen dose. Unfortunately, this distinction is almost impossible to make strictly, for it is always possible that the cells which respond to the initial experimental dose of an antigen have already been exposed to cross-reacting antigen in their environment. Nevertheless, in Table 8–2, where the evidence for cell specialization is summarized, the presumptive classification (B or T, primed or unprimed) of the cells involved is indicated.

Paul (1970) has recently reviewed the evidence for specialization of antigen-reactive cells, and this and other evidence is summarized in Table 8-2. It will be seen that there is considerable indirect evidence for the specialization of immunocompetent cells to a restricted range of specificities, immunoglobulin classes and subclasses, and allotypes. Perhaps the most dramatic demonstration is the phenomenon of *antibody-mediated suppression.* Passive immunization of fetal or neonatal animals with antibody directed against various immunoglobulin determinants severely suppresses the synthesis of antibodies with those determinants long after the immunizing antibody has been cleared from its circulation and presumably destroyed. This suppression is equivalent (in form at least) to the suppression of cells committed to antibody directed against endogenous antigens, the presumed explanation of tolerance to self. Like tolerance, antibody-mediated suppression is difficult to explain except by hypothesizing that cells are committed to the "unwanted" species of immunoglobulin, which can thus be effectively purged by suppressing those cells. Since the cells which synthesize the immunoglobulins are B cells, these suppression experiments demonstrate a specialization of antigen-reactive B cells, which have not undergone any experimental antigenic stimulation at the time of administration of suppressing antibody.

A similar degree of specialization of antigen-reactive cells is implied by a number of other ingenious experiments (Table 8–2). In *limiting dilution* experiments, one asks whether the capacity to produce two or more different antibody types (for instance, allotypes, classes, or specificities) appears independently in preparations in which the immunocompetent cells are so few that only a small number of the preparations produce either. Independence indicates that the original cells responsible for the two capacities are different. An example of *nonindependent* distribution is found in a paper by Shearer, Cudkowicz, and Priore (1969), who showed that under conditions of limiting T cells and excess B cells, the ability to produce IgG and IgM antibody appears simultaneously in most experiments; because T cells have been shown not to produce antibody, this lack of class specialization in them is not surprising.

In *affinity-killing* experiments, antigen-reactive cells are incubated with a highly radioactive "lethal antigen"; those cells which bind the antigen are killed as a result of the radiation, and it is found that this treatment ablates the ability of a population of cells to respond to that antigen, leaving unaffected its ability to respond to other antigens. The response to other antigens is therefore due predominantly to cells which do not bind the lethal antigen. *Response-killing* experiments are similarly designed to kill the cells capable of responding to one antigen and assaying the survivors for the ability to respond to another. In this case, however, responding cells are selectively killed by virtue of their high DNA synthesis.

The cells capable of making one type of antibody can also sometimes be physically separated from those able to make another. L'age-Stehr and Herzenberg found an adventitious difference in the buoyant density of memory cells capable of making IgG and IgM; and Abdou and Richter and Wigzell were able selectively to separate the cells capable of responding to one antigen by adsorbing them to affinity columns (see Table 8-2).

Simonsen (1967) demonstrated that as few as fifty lymphoid cells injected into an immunologically incompatible host chicken could regularly (five cases out of sixteen) initiate a graft-versus-host reaction. He reasoned that as high as 2 percent of these cells must therefore have been able to respond to the pertinent histocompatibility antigens. Similar results have been obtained from human histocompatibility differences when assayed by the mixed lymphocyte culture technique (Bach et al., 1969). Simonsen argued that if, as proposed in the clonal selection hypothesis, cells were stably committed to a single specificity, commitment of 2 percent of these cells to "that little corner of the immunological universe which is made of a single histocompatibility locus" is unreasonably large. I do not agree with this argument. One histocompatibility locus in the mouse corresponds to an average rate of recombination between its ends of about 0.5 percent; if one takes the average rate of recombination in this animal (Sager and Ryan, 1961) as applying to this small region, it is still large enough to accommodate thousands of

**Table 8–2.** Specialization of cells before the stage of antibody secretion.

Antibody-mediated suppression:

| Study | Species | Specificity suppressed |
| --- | --- | --- |
| Dubiski (1967a and b) | Rabbit | *b*5 ($C_\kappa$) in homozygotes<br>*b*4, *b*5 ($C_\kappa$) in heterozygotes |
| Appella et al. (1968); Mage, Young, and Reisfeld (1968); Vice, Hunt, and Dray (1969, 1970) | Rabbit | *b*5 ($C_\kappa$) in homozygotes |
| Dray (1962) | Rabbit | *b*5 ($C_\kappa$) in heterozygotes |
| Mage and Dray (1965, 1966) | Rabbit | *b*4, *b*5 ($C_\kappa$) in heterozygotes |
| Mage (1967); Mage, Young, and Dray (1967) | Rabbit | *b*4, *b*5 ($C_\kappa$) in heterozygotes<br>*a*1, *a*3 ($V_H$) in heterozygotes |
| David and Todd (1969) | Rabbit | *b*5 ($C_\kappa$) in homozygotes<br>*a*3 ($V_H$) in homozygotes |
| Tosi, Mage, and Dubiski (1970) | Rabbit | *a*1 ($V_H$) in heterozygotes* |
| Vice et al. (1970) | Rabbit | *a*2 ($V_H$) in homozygotes |
| Herzenberg, Minna, and Herzenberg (1967) | Mouse | *Ig*-1$^b$ ($C_{\gamma 2a}$) in homo- and heterozygotes |
| Ruffilli, Compere, and Baglioni (1970) | Mouse | Subgroup-specific $V_\kappa$ determinant |
| Kincade et al. (1970) | Chicken | IgM†, IgG |

Limiting dilution experiments:

| Study | Species | Independent distribution of– | Presumed cell-type |
| --- | --- | --- | --- |
| Osoba (1969) | Mouse | antisheep, antichicken RBC | $1^0$, T or B |
| Bosma and Weiler (1970); Weiler and Weiler (1968) | Mouse | $IgG_{2a}$ allotypes | $1^0$, B |
| Cudkowicz, Shearer, and Priore (1969) | Mouse | anti-RBC's | $1^0$, B |

Affinity-killing experiments:
Discussed by Ada (1970): $1^0$, B & T

Response-killing experiments:

| Study | Species | Specialization with respect to– | Presumed cell type |
|---|---|---|---|
| Zoschke and Bach (1970) | Man | various protein antigens | $1^0$, T |
| Zoschke and Bach (1971) | Man | histocompatibility antigens | $1^0$, T |
| Dutton and Mishell (1967) | Mouse | various RBC's | $1^0$, T or B |

Physical separation experiments:

| Study | Species | Separation method | Specialization with respect to– | Presumed cell type |
|---|---|---|---|---|
| L'age-Stehr and Herzenberg (1970) | Mouse | Buoyant density | IgM, IgG | $2^0$, B |
| Abdou and Richter (1969) | Rabbit | Affinity columns | sheep, horse RBC's | $1^0$, T and B |
| Wigzell (1970) | Mouse | Affinity columns | two proteins | $1^0$ and $2^0$, B |

*In these experiments suppression of *a*1 simultaneously suppressed the $C_\gamma$ allele inherited from the same parent (that is, on the same chromosome), but not the "trans" $C_\gamma$ allele on the other chromosome.

†Kincade et al. (1970) found that early suppression of IgM in the presence of the bursa simultaneously suppressed IgG.

structural genes, and thus thousands of different histocompatibility antigens. Moreover, by the arguments in Chapter 5, it is by no means necessary that the antibodies recognizing those specificities are devoted exclusively to them: they might recognize innumerable other foreign determinants as well, including some determined by other histocompatibility "alleles." In a similar histocompatibility system in humans, in which again very few cells were sufficient to produce a positive reaction, it was nonetheless demonstrated that the responding cells were specialized to a high degree for that particular histocompatibility difference (Zoschke and Bach, 1971).

While the experiments I have so far cited indicate that different immunological capabilities are largely segregated into different antigen-reactive cells, they cannot exclude a low incidence of cells harboring both capabilities simultaneously. In a number of experiments in which the presumed antigen receptors on the surface of potential antigen-reactive cells have been studied by the use of antisera directed against immunoglobulin determinants, some evidence has been obtained for cells with more than one type of receptor (see reviews by Greaves, 1970, and Sell, 1970). I am very skeptical of these conclusions, however. A recent immunofluorescent study by Pernis, Forni, and Amante (1970) of immunoglobulin determinants on the surface of normal lymphoid cells uncovered no cells in heterozygotes harboring two allelic allotypes on their surfaces, strongly indicating that allelic exclusion applies to these nonsecreting cells as well as to antibody-producing cells. A similar study by Davie and his co-workers (1971) did find a few cells (less than 3.4 percent) harboring two allelic allotypes, but the authors appear to be dubious about their interpretation as actual double-producing cells.

It should be noted that incomplete specificity is a frequent source of false double-labeling in specific immunological labeling experiments, whereas there is little technical criticism of experiments in which double-labeling is not observed. On the other hand, a theoretical objection can be made to experiments like those of Pernis et al. in which *normal* lymphocytes are examined: these might comprise almost entirely cells which have undergone some form of antigenic stimulation, and thus experiments on them do not necessarily answer the crucial question of whether cell specialization is antigen independent.

Pernis' group, in the same study, found that about 4 percent of the cells in the rabbit were positive for *a*11 or *a*12 (allotypes in the hinge region of $C_\gamma$—see Appendix B); about 35 percent were positive for *Ms*3 (allotype associated with the hinge region of $C_\mu$) and for class-specific IgM determinants detected by antiserum produced against whole IgM; and no cells labeled with antisera directed to the IgG Fc fragment. About half the cells failed to label with any of these reagents. These results suggest that many receptors on B cells are IgM. The failure to find IgG Fc determinants on any cells (even though hinge-region determinants were found) might be interpreted to mean that immunoglobulins as receptors assume a configuration in which the Fc domain is buried and inaccessible; a similar conclusion was reached by Greaves (1970).

The mere presence of immunoglobulin determinants on lymphoid cells does not prove that these are receptor molecules identical to the antibody those cells will produce. However, the ability of antisera against immunoglobulins to suppress the capacity of cells to respond to antigens (Greaves, 1970; Lesley and Dutton, 1970; Mason and Warner, 1970; Warner, Byrt, and Ada, 1970) supports these conclusions. All these workers agree that anti-L-chain antisera inhibit all types of reactions, both in T cells (Mason and Warner, 1970) and in B cells (Warner, Byrt, and Ada, 1970). Antisera to Fc determinants are far less potent (Greaves, 1970). Mason and Warner (1970) found that T cells could not be inhibited by antisera against any of the five mouse H-chain classes and subclasses, but Warner, Byrt, and Ada (1970) found that B cells could be inhibited by antisera against $\mu$ chains, but not by antisera against the other four classes and subclasses of mouse H chains. The apparent predominance of IgM over IgG as receptor molecules supports the finding of Pernis and his group. The T cells, in which no known H-chain determinants were found, might harbor a different class of immunoglobulin ("IgX"), which shares L chains with the known serum classes but has a different H chain.

Altogether, it seems to me that the evidence cited constitutes a striking corroboration of the clonal selection theory. It must be admitted, however, that because it is nearly impossible to prepare immunocompetent cells which can confidently be said not be have undergone some form of antigenic stimulation, the central question of whether cell specialization is antigen-dependent remains to be answered. There is, in fact, strong evidence suggesting that at least some lymphoid cells actually synthesize more than one molecular species of antibody, although at low levels. In some established human lymphoid cell lines, individual cells harbor more than one class of H chain; in other cell lines, a single class of H and L chain is found (see note a, Table 8-1). Because this double production has persisted through many cell divisions, the possibility of passive acquisition by one cell line of immunoglobulin produced by another can be excluded. But this cannot be taken as contrary to the clonal view, for it is not known whether these cells are at a functionally immunocompetent stage. In the differentiation leading to this stage, there may be a progressive loss of all but one H and L chain. It would indeed be very interesting to know the structure of the various immunoglobulins these cells produce, for they might give valuable clues to the early stages in the development of cells which eventually produce but a single antibody.

### Recommitment of Cells from IgM to IgG

Although the clonal selection theory supposes that one cell is precommitted to one immunoglobulin, it does not require that this commitment be irreversible. One type of recommitment has in fact been well established. During many immune responses there is a shift in serum antibody from IgM to IgG (see, for example, Bauer, Mathies, and Stavitsky, 1963). A number of lines of evidence suggest that this switch reflects a subcellular event which converts some cells committed to IgM to the production of IgG. First, the case of the myeloma patient Til to be discussed in Chapter 9 is a direct demonstration that such a switch can occur. Secondly, Oudin and Michel (1969) have shown that the IgM and IgG antibody produced in a single rabbit share common idiotypic (that is, presumably V-region specific) determinants. This strongly suggests that the IgM and IgG derive from a common stem of cells; for if they derive from independent cells, they would be expected to display different V-region determinants, as do antibodies produced by different rabbits. Thirdly, cells producing (or at least harboring) both IgG and IgM have been described during the period of the switch from IgM to IgG in the serum (for example, Cosenza and Nordin, 1970), in contrast to the general rule that only one of several alternative immunoglobulin classes can be found in any one cell. The reported frequency of these double producers appears to be only a few percent, however, and it is argued that this low percentage rules out the hypothesis that IgG producers all derive from IgM producers. The argument, however, rests on implicit assumptions about how long the switch period in individual cells lasts, and thus what proportion of cells can be expected at any one time to be caught in the intermediate period in which both IgG and IgM can be detected. Fourthly, Kincade and his fellow workers (1970) have shown that early administration of anti-IgM antibody to chickens suppresses both IgM and IgG production, while anti-IgG suppresses only IgG. These lines of evidence, taken together, constitute a strong case for a recommitment of cells from IgM to IgG of the same specificity.

While this recommitment itself seems established beyond cavil, it is by no means certain whether it occurs before or after antigenic stimulation. The above-mentioned results of Warner, Byrt, and Ada (1970) and Pernis, Forni, and Amante (1970), suggesting that most receptors on unprimed antigen-reactive B cells may be IgM, would support the contention that the switch occurs after antigenic stimulation and thus might explain the corresponding switch in serum antibody class during the immune response. However, this is indirect and uncertain evidence and is contradicted to some extent by Kincade's suppression studies in chickens; he found that the ability of anti-IgM antibody to suppress IgG as well was dependent on the bursa and could be lost soon after hatching, presumably before most antigenic stimulation occurred in these animals. The results suggest, instead, that cells committed to IgM become recommitted to IgG under the influence of the bursa and independently of antigen. It is, of course, possible that both antigen and the bursa are required, for the supposition that few cells have undergone antigenic stimulation by the time of hatching might be quite wrong.

This one relatively well-established case of multipotency of single antibody-producing cells seems to be the exception that proves the rule of cellular specialization. Wherever it has been tested, it has been found that the IgM and IgG produced by a single cell line share everything in common but their $C_H$ regions. Thus the heritable commitment of the cell line to the original $V_H$, $C_L$, and $V_L$ regions remains apparent even after a recommitment to a new class of antibodies with markedly different properties.

## Information Transfer Experiments

In addition to the putative "helper" function of T cells in the antibody response, there is another aspect of cell cooperation in the immune system which, if some of its implications are accepted, undermines the whole clonal selection view, as well as many of the other concepts that have been treated in this book. Researchers (Adler, Fishman, and Dray, 1966; Bell and Dray, 1969, 1971) have shown that RNA extracted from peritoneal cells (largely macrophages) of an immunized rabbit can induce a culture of spleen cells of another rabbit to produce antibody. There is good evidence that some of this RNA carries genetic information; when the RNA donor and spleen cell donor carry different allotypes (alleles) for immunoglobulin genetic markers, it is found that in the early in vitro response much of the antibody (which at that time consists mostly of IgM) carries both $V_H$- (*a* locus) and κ-chain (*b* locus) markers characteristic of the RNA donor. The experiments seem to be well controlled, even if they deal with rather marginal immune responses, and the reported results of fairly independent experiments agree with the above conclusion: that peritoneal cell RNA can transfer structural information for immunoglobulins to spleen cells.

If we accept the informational nature of the RNA extract, there can be two possible interpretations. On the one hand, the RNA may be normal immunoglobulin messenger RNA (perhaps, for some reason, particularly easily obtained from peritoneal cells even though these are not presumed to be major producers of antibody), and the spleen cells may serve as translation factories in which this RNA is used to synthesize antibodies which the spleen cells otherwise would not make. According to this view, the phenomenon described by these authors does not reflect any physiological process but is merely a virtuoso trick in protein synthesis. This view is comforting in that it violates none of the basic tenets of conventional immunology. However, it is difficult to believe that the extracts used in these experiments contain enough messenger RNA corresponding to the antibody specificity in question, and of such longevity under the conditions of the experiment, that a detectable immune response peaking three days after the administration of RNA could result.

These limitations have led to the other interpretation, which is that the informational RNA present in the original extract is replicated inside the recipient cells. This putative replication is a very complex addition to the picture, and if it occurs in this experimental situation it is difficult to avoid the conclusion that the apparatus is involved in normal immune responses. It is conceivable that transfer of informational nucleic acid to other cells, and subsequent replication in them, might greatly enhance an otherwise slow immune response limited by the synthetic and proliferative capacities of a few original responsive cells. The RNA viruses do provide some precedent for this view, in that RNA-dependent RNA polymerases or RNA-dependent DNA polymerases are found in cells infected by them.

Whether or not such an information transfer is a normal physiological process is subject to experimental test. In a "tetraparental" chimera formed from two strains of mouse which differ at an immunoglobulin genetic marker locus and at some other marker locus (such as H2), it can be ascertained if any cells producing antibody with the immunoglobulin marker coded by one parent carries the other marker specified by the other parent, as predicted if information transfer between cells occurs at all regularly. Preliminary experiments indicate that such cells are rare but the possibility of their existence is not completely eliminated (T. Wegman and L. A. Herzenberg, personal communication).

It is difficult to make much sense of an immune response involving information transfer. Why should this information involve the macrophage (or other peritoneal cell), which neither makes antibody nor (so far as anyone knows) responds specifically to antigen? How is the specialization of individual antibody-producing cells explained if they are subject to the receipt of informational nucleic acid from every quarter? The complications of such a scheme seem to outweigh any conceivable advantage. Consequently, I have ignored these rather disturbing experiments until they are confirmed by a plausible theory.

I do not think that any of the experimental evidence bearing on the clonal selection theory, either for or against, can be taken as entirely conclusive. Once one accepts the a priori possibility of subcellular selection (which I am unwilling to do), several explanations for the observed specialization of most antibody-producing cells can be advanced (for example, a positive feedback locking in production of the antibody which happens to be triggered first). Nevertheless, the evidence is mostly compatible with clonal selection, and until absolutely incontrovertible proof to the contrary is obtained, I think the three fundamental postulates of Burnet's theory must be retained, for the denial of any one of them would imply the existence of mechanisms far more implausible than the ad hoc modifications required to reconcile the theory with a few pieces of awkward evidence. The best evidence for the clonal selection theory, in short, is the phenomenon of adaptive immunity itself.

As it was originally advanced by Burnet, the clonal selection theory was coupled to a somatic mutation mechanism of generating antibody diversity. He implicitly assumed that the number of mutating genes was severely limited—perhaps to only one cistron—and the restriction of a single cell to one specificity was a more or less natural outcome of this restriction of structural genes. Clearly the problem of cell specialization (whether or not it is independent of antigenic stimulation) is not as simple as that. Table 8-3 lists all the human germline genes that have been demonstrated; if the germline theory is correct, the number of V genes may in fact number in the thousands. Each diploid cell presumably has two of each gene. Yet of these many genes, only four are expressed in any one cell at any one time;

**Table 8–3.** Well-established germline genes in humans. Representative sequences deriving from each of the deduced V genes are listed in parentheses. $V_{\kappa}I$, $V_{\kappa}II$, and all the C genes have been shown to be nonallelic to one another.

| V genes | C genes |
|---|---|
| κ family | |
| Ia (Ou, Hau) | κ |
| Ib (Bel, Au, Ag, Roy) | |
| Ic (Dav, Fin) | |
| Id (Eu, HBJ4) | |
| II (Ti, Rad, B6, Fr4) | |
| III (Cum, Tew, Mil) | |
| λ family | |
| III (X, Bau, Kern) | λarg |
| V (Sh) | λlys |
| I, IV (Ha, New) | |
| II (Vil, Bo) | |
| H family | |
| I (Eu) | μ |
| II (Daw, Cor, Ou, He) | α1 |
| III (Nie) | α2 |
| | γ1 |
| | γ2 |
| | γ3 |
| | γ4 |
| | δ |
| | ε |

their selective expression is certainly far from a trivial and unimportant problem.

The next chapter will discuss the evidence that V and C genes are not linked on a one-to-one basis in the genome, but must be linked somatically in order to be expressed as whole immunoglobulin polypeptide chains. It will be proposed that the reason only one L chain and one H chain is expressed per cell is that only one pair of $V_L$ and $C_L$ genes and one pair of $V_H$ and $C_H$ genes are normally joined.

**Chapter 9**

# The Joining of V and C Regions: Theories of Cell Commitment

The fact that the genealogical relationships apparent at all positions of V regions conform topologically to one another within the limits of expectation implies that V genes have evolved as single genetic units and have not recombined significantly. In strong contrast to this result, genealogies reconstructed for the V and C regions of the same polypeptide chains do not conform. Figure 9-1 shows the contrasting genealogical relationships between the V and C regions of 52 immunoglobulin polypeptide chains. The evidence is compatible with the notion that—with the restriction that only $V_\kappa$ can be associated with $C_\kappa$, $V_\lambda$ with $C_\lambda$, and $V_H$ with $C_H$—V and C regions are associated at random. The apparently random association of V and C regions within the same family implies that the corresponding genes cannot be associated in the germline on a one-to-one basis. These arguments are quite independent of the difficult question of whether C regions are encoded by unique germline genes, but are based instead entirely on genealogical reasoning. The case of the human myeloma patient Til who produces both an IgG and an IgM with apparently identical $V_H$ regions (to be discussed later) strongly confirms the supposition that a single V gene can be associated with any C gene in the same family.

Although the evolution of V and C regions is nonparallel when examined in detail, overall the V and C regions of the mammals have evolved in parallel to form three independent gene families, one each for $\kappa$, $\lambda$, and H chains. A cursory examination of the V regions in Tables 2-2A, 2-3A, and 2-4A shows that $V_\kappa$ and $V_\lambda$ regions are much more closely allied to one another than either family is to the $V_H$ regions; these relationships parallel those of the corresponding C regions, since $C_\kappa$ and $C_\lambda$ evidently diverged recently in comparison to the divergence of their common ancestor from the ancestral $C_H$ gene (Chapter 4). Within each of these families, the V and C genes have undergone independent duplications to produce the various V-region subgroups and C-region classes and subclasses. The H and L chains, in this view, are the result of the duplication of an entire original immunoglobulin gene family; and the L-chain gene family more recently duplicated in its entirety to produce the present mammalian $\kappa$ and $\lambda$ families. While genes from different families are never associated in the

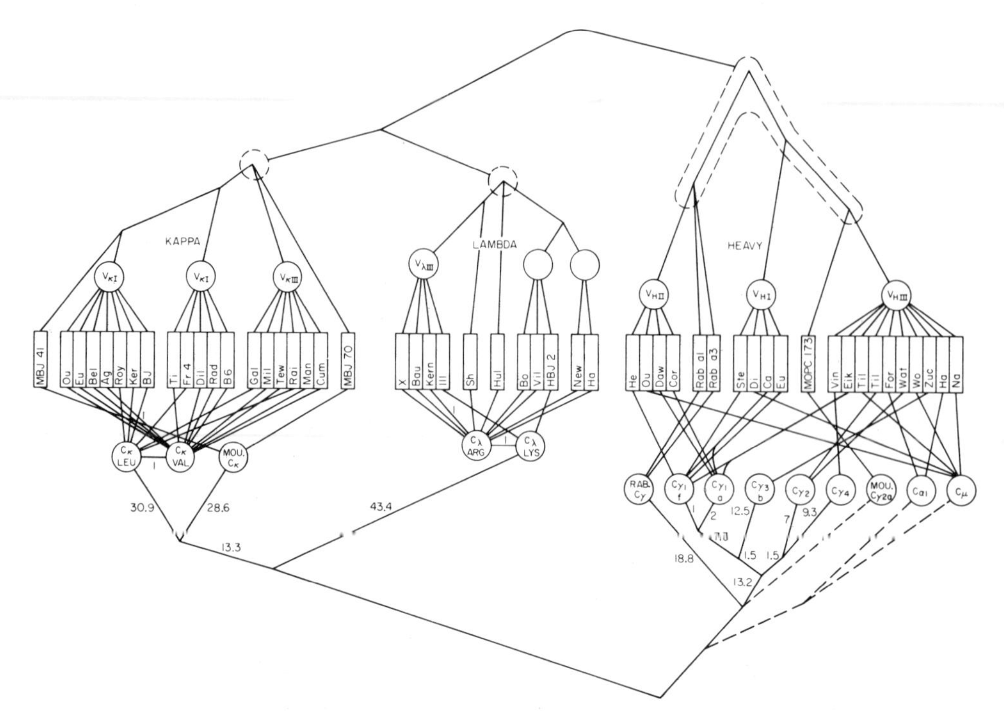


**Figure 9–1.** Nonparallel evolution between V and C regions of a number of immunoglobulin sequences. This result precludes any stable one-for-one association of V and C genes in the germline. Taken from Smith, Hood, and Fitch (1971).

same polypeptide chains, V and C regions from the same family are joined to each other freely and apparently at random. The chromosomal arrangements of the immunoglobulin genes support this view; genes from the same family are linked to one another, while those from different families are unlinked (Fig. 9–2).

Hamers and Hamers-Casterman (1967) showed that in a doubly heterozygous rabbit with allelic markers at both the $V_H$ and $C_\gamma$ loci, only parental combinations of the V- and C-region allelic forms were expressed in individual γ chains, implying that at least in this case only V and C regions coded by the same chromosome are joined; the predominance of intrachromosomal over interchromosomal joining has since been confirmed in the rabbit with different $C_\gamma$-region allelic markers (Prahl et al., 1970; Tosi, Mage, and Dubiski, 1970).

The joining of V and C regions might take place at one of three stages: the peptide level, the messenger RNA level, or the DNA level. The failure

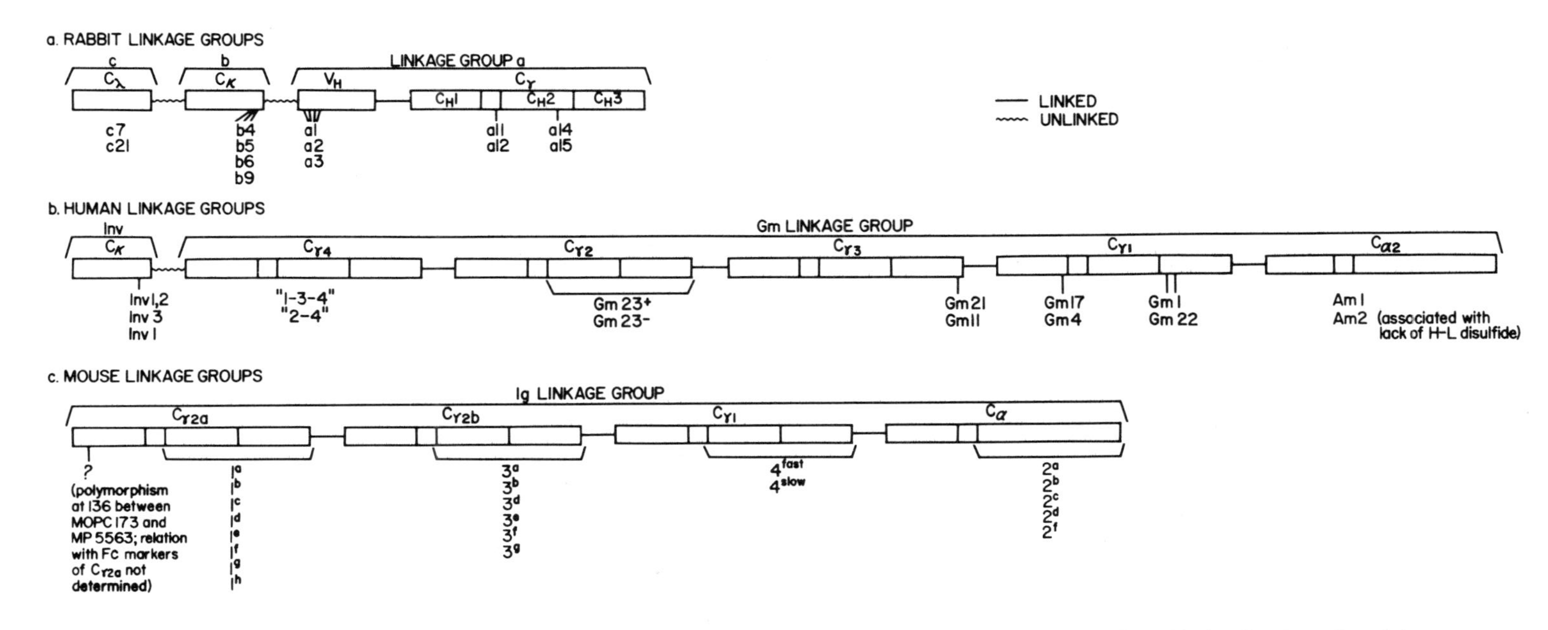


**Figure 9–2.** A genetic map of immunoglobulin structural genes. Major allelic allotypes for each gene are indicated below it. The order of the human $C_{\gamma2}$, $C_{\gamma3}$, and $C_{\gamma1}$ genes is taken from Appendix B; the order of the genes in the rabbit *a* and mouse *Ig* linkage groups is not known. If the DNA network hypothesis is correct, these orders are not meaningful.

to demonstrate two growing points in nascent H and L chains (Fleischman, 1967; Knopf, Parkhouse, and Lennox, 1967; Lennox et al., 1967) argues against peptide-level joining. Moreover, as Terry and Ohms (1970) have pointed out, the structural gene for the protein Zuc (see Tables 2-4A and 2-4B), in which a long sequence spanning the V-C junction is deleted, must represent a V-C joining at the DNA level. Otherwise one would have to postulate that messenger RNA or polypeptide half-chains are joined despite the deletion of precisely those parts of the sequences which are normally joined. A case (and some rather indirect evidence) for peptide-level joining has been put forth by Schubert and Cohn (1970), but until more compelling facts are advanced to back this hypothesis, which is reconciled with most of the evidence only by some rather special assumptions, I shall assume with most immunologists that V and C genes are linked at the level of DNA.

Several lines of evidence suggest that the mechanism by which V and C are joined is intimately related to the mechanism by which a cell becomes heritably committed to the synthesis of a single molecular species of immunoglobulin. Chapter 8 presented evidence that in the course of the immune response cells become recommitted from the production of IgM to the synthesis of IgG. This recommitment seems to be associated with a rejoining of the $V_H$ region from the $C_\mu$ region to the $C_\gamma$ region. Thus, in the human myeloma patient Til, there are two cell populations—one producing IgM, the other $IgG_2$. The $\kappa$ chains (Wang et al., 1969) and the $V_H$ regions (see Table 2-4A) of the two proteins appear to have identical amino acid sequences, and further presumptive evidence for sequence identity comes from the many "idiotypic" (V-region specific) antigenic determinants shared by the two proteins (Levin et al., 1971; Wang et al., 1970b). Similarly, the idiotypic determinants present on the IgM and IgG antibody produced by a single rabbit are very similar, even though antibodies produced to the same antigen in different rabbits share few idiotypic determinants (Oudin and Michel, 1969). Thus, during the recommitment of a cell from one H-chain class to another, there appears to be a concomitant rejoining of the $V_H$ and $C_H$ regions. This suggests that commitment and joining are related; the only alternative is that cells become committed and recommitted to V and C genes independently of one another, which seems an unattractive complication of the commitment mechanism.

### Theories of Joining and Commitment

Two main accounts of this intrachromosomal, DNA-level joining of V and C genes have been advanced. The translocation model (Fig. 9-3) proposed by Dreyer and Bennett (1965) and others (Dreyer and Gray, 1968; Gally and Edelman, 1970) envisions the multiple V and C genes in one family to be in tandem array. One V gene is, by means of one or more DNA breakage and reunion (crossing-over) events, placed next to one of the C genes. This joining must be very precise to explain why so many chains have identical lengths near the junction of their V and C regions. A

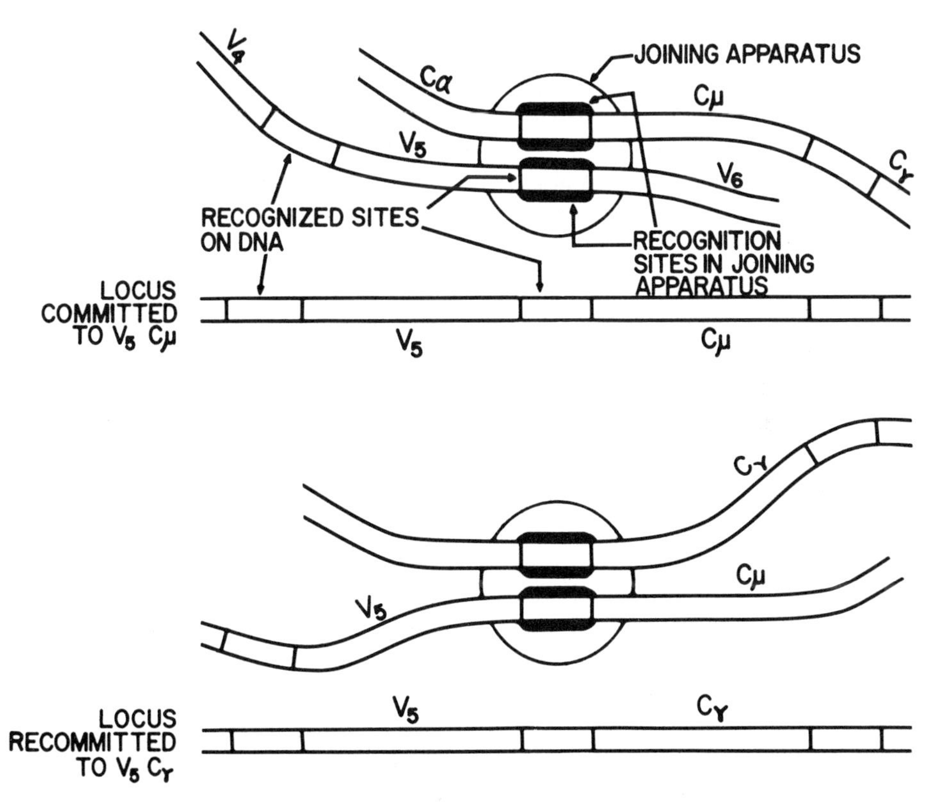


**Figure 9-3.** Translocation model for V-C joining. The joining event is unspecified: it could be a single crossover within a single DNA molecule, an insertion of an episomal V or C gene, or even insertion of a replicated copy of the V or C gene.

specific apparatus is usually invoked to perform this gene translocation, perhaps similar to the specific enzymes *int* and *xis* which integrate and excise λ-phage DNA into and out of the *E. coli* genome.

Such a translocation event would precommit the gene family in which it occurred to the expression of the particular pair of V and C genes joined. This commitment, which would be an integral feature of the DNA sequence, would be naturally inherited by the clone of cells to which that gene family was transmitted. If some subsidiary hypothesis is formulated to explain why only one of the two H-chain families and one of the four L-chain families in each cell undergoes a translocation, then the heritable commitment of the cell to a single molecular species of antibody is explained.

According to this hypothesis, recommitment would be essentially the same as the original translocation event. The $V_H$ gene originally associated with, say, $C_\mu$ would simply be translocated to a new $C_H$ gene (say $C_\gamma$), and the cell and its progeny would be committed to a new immunoglobulin with only the $C_H$ region changed. Any of the other three regions of the molecule might be switched in the same way, although there is no evidence that these theoretically similar processes occur. In general, a recommitment which switches more than one of the four regions expressed would not be expected, since this would probably require more than one translocational event.

Smithies (1970) has proposed a *DNA network hypothesis*, shown in Fig. 9-4, in which the multiple genes in each family are in lateral rather than tandem array. The creation of new branches is described in the above reference: essentially, a single crossover in the two double-stranded branches produced transiently during DNA replication introduces a twist in them which allows the entire branched structure to be replicated and the replicated copies separated, with a single simple set of rules for joining the new single-stranded ends produced at each bifurcation. Additional assumptions are required to allow DNA and RNA polymerase to pass such forks in the resulting DNA network.

Thus in Smithies' model all V genes are simultaneously colinear with all the C genes in the same family: there is no necessity for any physical joining of genes as by the translocation hypothesis. However, some mechanism must single out which of the several V genes will be expressed with which of the several C genes in the immunoglobulin chain produced by each network. Smithies proposes that two-position protein "switches" are attached to each bifurcation in the network. Depending on the setting of the switch at that bifurcation, RNA polymerase will follow one of the two possible paths at each fork. A completely random attachment of these switches will specify a single path for RNA polymerase through the network, thus a single messenger RNA sequence. It is also possible that the setting of some of the switches is under strict physiological control.

In order to explain the heritability of the commitment of the gene family to a single polypeptide, Smithies must hypothesize that during DNA replication, switches are attached to the daughter copies in settings identical to those on the parent DNA molecule. Although a general scheme for this switch-setting replication can be envisioned based on dimerization of the switch proteins, it seems a rather contrived aspect of the theory. Under the translocation hypothesis, by contrast, commitment is inherent in the DNA structure itself and thus naturally heritable.

Like the translocation hypothesis, Smithies' hypothesis seems to require some subsidiary assumption to explain why only two of the six gene families presumably present in each cell are active in immunoglobulin production.

Recommitment of a gene family to a new polypeptide is easily accommodated in Smithies' model: a resetting of any one of the switches in the transcribed path will specify a new path with either a new V region or a new C region. Since the DNA network is proposed to narrow down to a single branch between V and C genes (see Fig. 9-4), simultaneous switching of both V and C regions is not possible.

A rather complex chromosomal arrangement is required under the DNA network hypothesis to produce a protein such as Zuc (see Tables 2-4A and 2-4B), whose deletion spans the junction of V and C regions. Smithies and his associates (1970) have suggested that the major deletions observed so frequently among the immunoglobulin chains are a natural outcome of interchromosomal crossing over between homologous branches in the proposed DNA network. If the crossed-over chromosomes are to segregate, the two DNA branches in which the crossover occurred must break, and the rejoining of the broken ends within each chromosome can lead to each of the patterns of deletion proteins that have been observed. The explanation for Zuc, however, requires two independent crossover events, to generate broken ends both on a V-gene branch and on a C-gene branch. Smithies' account of the major deletions differs from the corresponding account on the tandem model in that it invokes only a simple crossover to generate broken ends in the DNA. Many other mechanisms for generating these breaks can be envisioned, so that these deletions cannot be taken as very strong evidence in favor of the DNA network hypothesis.

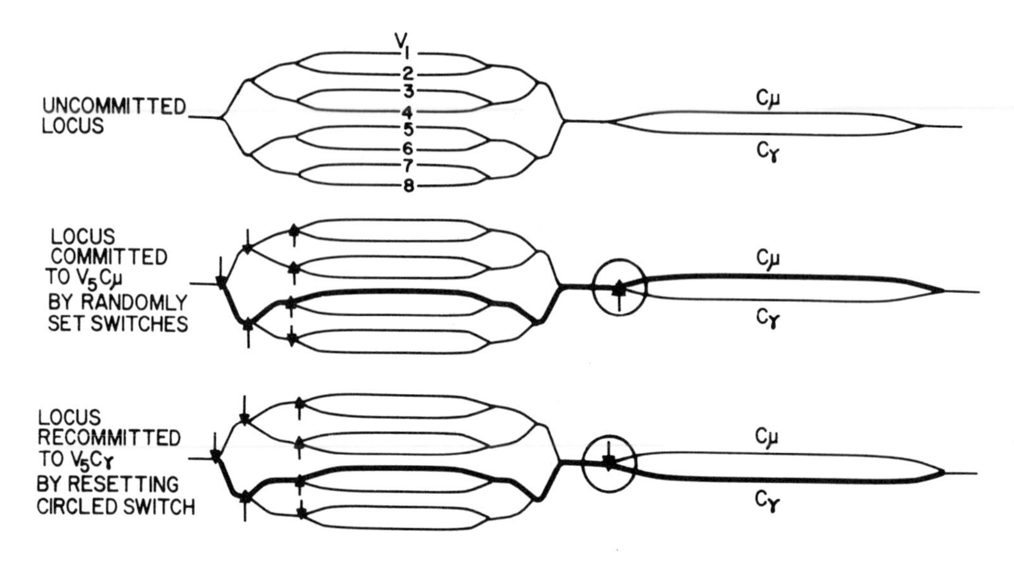


**Figure 9-4.** The DNA network model of Smithies (1970).

## The Pattern of Variation at the V-C Junction of κ Chains

A clue to the mechanism of joining might be obtained by examining the pattern of amino acid sequence variation at the junction of V and C regions. As mentioned in Chapter 6, there is apparent genealogical scrambling at the end of $V_\kappa$ regions which is not observed elsewhere in the chains. Figure 9-5 illustrates this pattern. The human κ proteins are listed, along with their corresponding messenger RNA nucleotide sequences for positions 100 to 108; nucleotide positions which are unknown or which do not vary are left blank. At the left of the figure is indicated the single descent consistently suggested by the pattern of variation at positions 1 to 99 (see Chapter 6). Lines separate the three major subgroups, $V_{\kappa I}$, $V_{\kappa II}$, and $V_{\kappa III}$, into which these sequences fall, and the two "sub-subgroups," *a* and *b*, into which the first of these can be divided. At positions 100.1, 100.2, 103.2, and 104.1 there are two alternative nucleotides which occur in at least three sequences each; at none of these positions do the alternatives present correlate with the subgroups or genealogical relationships evident in the first 99 positions.

If we compare this peculiar pattern of variation at positions 100 to 108 of human κ chains with some simple expectations of these two theories,

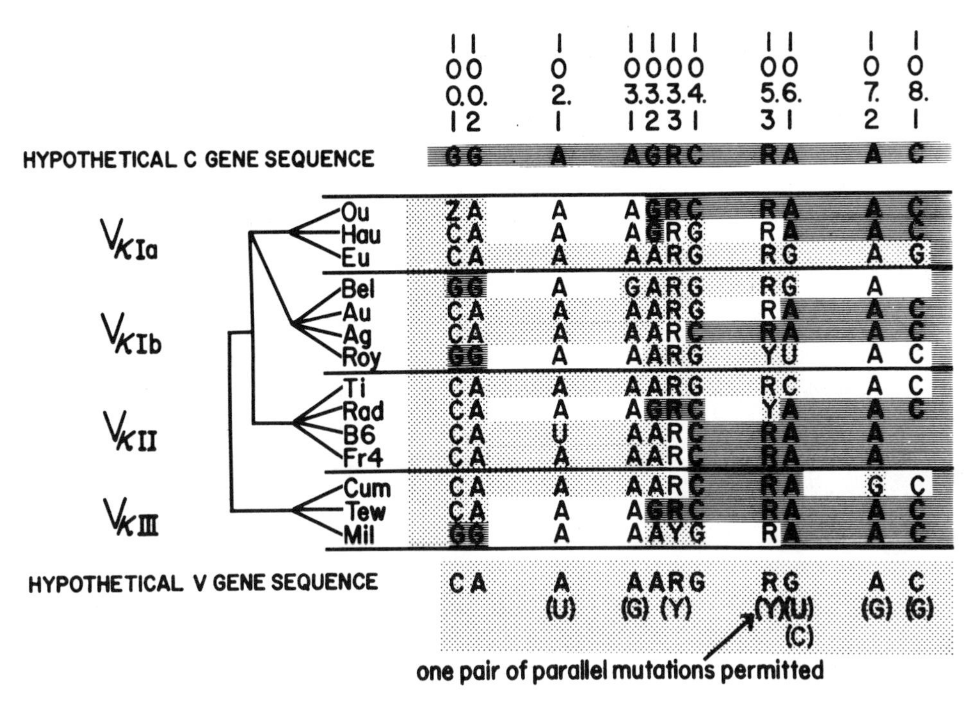


**Figure 9-5.** Pattern of variation at the end of human $V_\kappa$ regions.

the pattern is consistent with either theory, but neither explains it very economically.

If according to the translocation model the breakage and reunion which joins V and C genes can occur anywhere within a short region where the two overlap, the true junction between V- and C-determined sequence might occur at different positions within the overlap region in different sequences. This might in principle lead to a pattern of variation which does not correlate with that expressed in the remainder of the V regions. Figure 9-5 illustrates this possibility. At each of the four above-mentioned nucleotide positions, one of the alternatives was assigned to the V genes and one to the C gene. A systematic examination of all possible assignments determined those in which the fewest crossovers were required. The remaining variable positions were then used to arrive at a single most parsimonious solution, in which the required crossovers were minimized and in which no variation was permitted in the C gene. (A single pair of parallel mutations was permitted at position 105.3 of the Roy and Rad V genes, for otherwise six additional crossovers would have been required.)

The final assignment is shown in Fig. 9-5; in that figure consecutive V-assigned positions have been linked by stippling and consecutive C-assigned positions by parellel lining. The positions N-terminal to 100 are, of course, assigned to V genes, while those C-terminal to 108 are assigned to the C gene. Eight of the fourteen chains included can, in this assignment, be accounted for by a single crossover; the remaining six require three (there must, of course, be an odd number).

This is not an unreasonably complicated account of the facts, but it shows a few disturbing features. First, five of the twenty-six crossovers occurred between positions 103.2 and 104.1, which are separated by a single nucleotide position. Second, while most of the V- and C-gene assignments are satisfactory in that they often occur with alternatives of like assignment in consecutive runs, the assignment of the GG sequence at positions 100.1 and 100.2 to the C gene is most unsatisfactory in that these alternatives are not linked to other C-assigned alternatives in any of the three chains in which they occur. Third, all of the triple crossovers can be attributed either to this GG sequence or to a single nucleotide with the "wrong" assignment which interrupts an otherwise unbroken run of nucleotides with the same assignment. These three facts imply a markedly nonrandom pattern of crossing over and necessitate an additional assumption: for example, that the "joining apparatus" has different preferences for the various nucleotides in the overlapping sequences it joins.

The branched network hypothesis of Smithies is also able to explain variation which is not linked to that occurring elsewhere in the chains; this lack of linkage would be a consequence of bifurcations occurring outside already existing ones. Indeed, any pattern of variation at all can be accounted for by such a network by postulating a series of bifurcations occurring successively outside one another. Such an account becomes plausible only if the resulting network is relatively simple.

The branched network in Fig. 9-6 is the simplest one compatible with the subgroup assignments and the pattern of variation at positions 100.1, 100.2, 103.2, and 104.1; no parallel mutations were permitted at these positions, as in the translocation hypothesis above. Included in parentheses in this network are minor variants which have been observed at other nucleotide positions. Unless these are to be attributed to allelic or somatic mutation, they too will introduce branches in the network, in which the entire branched structure to their left is duplicated. The roman numerals at the bottom of the diagram indicate the order in which the bifurcations had to occur in order to result in this network. This cannot, in my opinion,

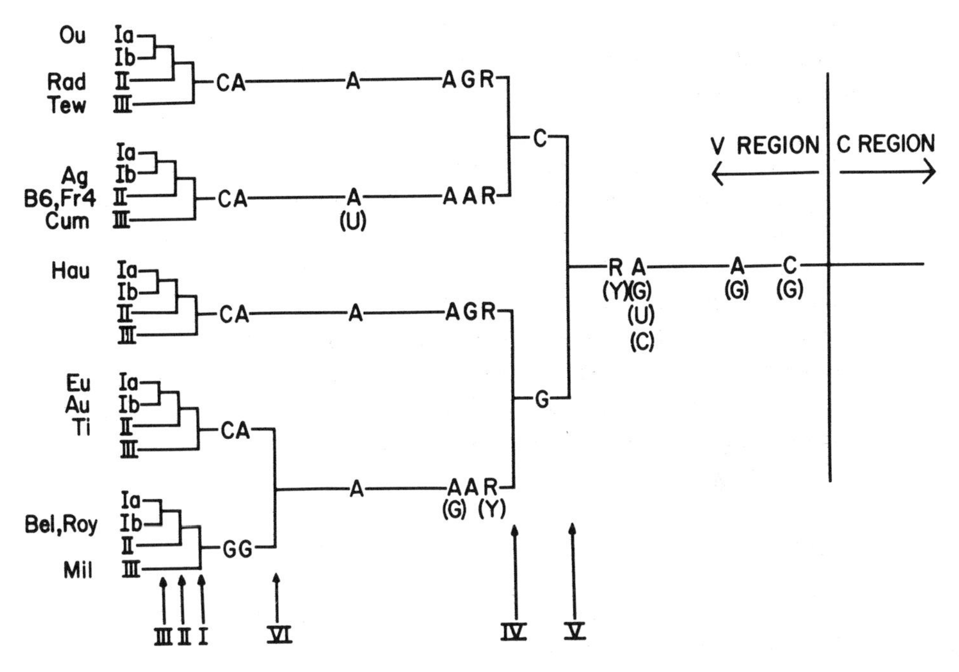


**Figure 9-6.** A minimum branched DNA network for the end of human $V_\kappa$ regions.

be considered a very elegant explanation of the data. The resulting network is complicated, and the degree of complication increases if the parenthesized variants are considered. There do not, in short, seem to be those restrictions on the number and linkage relationships of the variations in this region which would be expected if the network here had few branches. Neither the branched network nor the translocation theory, then, seems to explain the peculiar variation in this portion of $\kappa$ chains very elegantly.

Additional data about the junction between the V and C regions of H chains would be very useful, for in this family, in marked contrast to the $\kappa$ and $\lambda$ families, the various C genes differ markedly from one another and there is some hope of being able to sort out those residues which are coded by V and C genes. A comparison between the junction regions of a large number of H chains whose V regions belong to the same subgroup would be particularly useful.

Genetic crosses involving the closely linked $C_H$ genes in mouse, man, and rabbit, however, would provide the most critical evidence discriminating between a tandem and a lateral arrangement of genes. If the lateral arrangement obtains, one should find recombination patterns which are inconsistent with a single tandem arrangement. So far, not enough recombinants have been demonstrated with sufficient certainty to permit a conclusion (see Appendix B).

Selective expression of germline genes is, of course, not peculiar to the immune system; in the next section I shall propose that most cellular differentiation has the same molecular basis—programmed gene translocation—which is one attractive explanation of the specialization of antibody-producing cells; if that view is correct, the controlled joining mechanism hypothesized here for immunocytes can be seen simply as one particular adaptation of a very ancient mechanism which arose in evolution with the first highly differentiated organisms. Smithies (1970) has pointed out that the branched DNA network hypothesis, too, provides a useful model for cell differentiation.

**Programmed Gene Translocations and Differentiation**

The diversity of antibodies calls attention to itself, and has become one of the most interesting problems in modern biochemistry, for two reasons: the first is its appearance in a single physiological system, which becomes thereby markedly atypical; the second is its apparently preadaptive nature, which creates special problems for its adaptive control—solved, as we have seen, by the clonal selection theory.

If we ignore these two special features, antibodies become just one more example of adaptively controlled diversity, which is the hallmark of the highly differentiated organism. I see no reason why the same basic mechanism—clonal specialization—which seems necessary in the case of antibodies, may not account for the adaptive control of these other examples of diversity as well. This clonal commitment is called *determination* in modern developmental theory, and there is considerable evidence (Brown and Dawid, 1969) that cells are indeed committed in a heritably stable fashion to certain fates during ontogeny.

In a general way, one might think of the alternative to clonal determination as a theory analogous to subcellular selection in immunity. Such a theory of development certainly cannot be ruled out on the same evolutionary grounds used to rule out its counterpart for immunity, for there is every reason to suppose that most diversity has survived in nature precisely because each variant has been individually selected for a certain specialized function, and the same selection can be invoked to account for coordinate evolution of the subsidiary recognition sites necessary for its subcellular control. That this in fact happens—and often—in evolution is elegantly demonstrated by the intricate mechanisms by which *E. coli* achieves the adaptive expression of many of its structural genes.

There is profit, nonetheless, in considering the view that this precedent may have been abandoned in many cases in the evolution of higher organisms. While a general coherent picture of development does not exist from any point of view, clonal determination at least allows some simplification of the problem. If the subcellular theory is to be held, the decision to express or not to express a given physiological system apparently must be made more or less independently for *each cell in the organism.* If the clonal view of development is taken, on the other hand, a given developmental decision, made once in a single cell, applies heritably to its clone of progeny. If we keep in mind that no such heritable decision need be irreversible, it can be concluded that clonal commitment allows in theory for development to occur with far fewer biochemical decisions (whatever their nature) than would otherwise be necessary.

A more or less equivalent simplification of the subcellular theory of differentiation can be brought about by hypothesizing positive feedback controls, which would have the effect of locking in the differentiated state of a cell and its progeny long after the original stimulus was removed. Alternatively, switches controlling gene expression might be replicated in the same settings during cell division. In this sense, a commitment need not be *genetically inherited* to be clonal and long-lasting. But to me this is an unattractive complication of the picture. The reason that many biologists have been willing to consider it has, I think, been that the mechanism of a genetically heritable clonal commitment has been even more difficult to conceive, so that subcellular control, well precedented in the bacteria, has often been the preferred working hypothesis. I have tried here to suggest, in accord with the ideas of Dreyer and his co-workers (Dreyer and Bennett, 1965; Dreyer and Gray, 1968; Dreyer, Gray, and Hood, 1967), that gene translocation may reasonably explain clonal commitment in the immune system, and I suggest, also in accord with the views of these authors (Dreyer, 1970), that the same mechanism may serve as a model for clonal commitment in development.

The genome of a higher organism, in this view, would comprise in part a series of genetic modules of greater or lesser complexity, each consisting in a large or small number of cistrons (or perhaps, as in the case of antibodies, half-cistrons). Cell determination, then, would consist in the programmed translocation of these modules to appropriate positions in the genome, in the same manner that an engineer achieves the construction of a great variety of complex machines by the simple operation of plugging in appropriate modules from a relatively limited modular library. Such a translocation would be an integral part of the DNA of the affected cell and would be faithfully passed on to its progeny in the normal course of replication, until another translocation reversed or otherwise altered the previous commitment. A module potentiated by such a translocation might come under the control of regulating mechanisms acting on the "constant" region DNA into which it is plugged. The same regulating apparatus could then be used to specifically control the activity of different genetic modules (in other words, genes) in different determined cells.

It is not impossible to test a translocational theory of development. One might, for example, show that a given gene (identified with the aid of a genetic marker or by RNA-DNA hybridization) regularly changes its position between two identifiable components of the genome. And if it should happen that antibodies are not the only case in which half-cistrons are regularly translocated, gene translocation might be demonstrable in other systems by the same methods which so dramatically suggested its occurrence in immunity.

If an *E. coli* cell is immersed in a solution of lactose, it responds by becoming specialized for the production of the enzymes appropriate for the use of that metabolite, and this may in one sense be taken as a model for "differentiation." In a deeper sense, however, this model is essentially deficient for the supposedly analogous process in higher organisms. In the case of microbial "differentiation" it is the experimenter himself who programs the differentiation by controlling the environment of his charges. It cannot be reasonably maintained that the outside environment of a developing multicellular organism directly programs the adaptive specialization of its individual cells: the programming stimuli (if any) must originate within the developing organism itself and therefore must themselves be programmed. The mechanism of this internal "programmed programming" is a basic conceptual problem of development. The translocational theory supported here is silent on the fundamental difficulty of the programmed programming of biochemical decisions in development. Nevertheless, as a clonal theory it ameliorates the difficulty somewhat by radically reducing the number of these decisions, and in genetic translocation it suggests a relatively simple and specific molecular basis for their content though not their cause.

The immune system has often served as a model of cell differentiation. Investigators who think of it in this way have often focused on the highly visible cellular events which follow antigenic stimulation—the cell proliferation and antibody production which are the obvious signs of an immune response. But I suggest that it is in the recondite events which precommit a cell to the production of a single antibody specificity and thus allow antibodies to be adaptively expressed that the value of the immune system to a general understanding of development really lies.

# Appendixes References Index

# Appendix A

## Properties and Provenances of Immunoglobulin Sequences

**Table A-1.** Properties and provenances of $\kappa$ chains.

| Protein | Species | Source[a] | Subgroup | Allotype | References |
|---|---|---|---|---|---|
| Ag | Human | MP | I | Val | Titani, Shinoda, and Putnam (1969) |
| Ale | Human | MP | I | | Milstein, Milstein and Feinstein (1969) |
| Au | Human | MP | I | Val | Schiechl and Hilschmann (1971) |
| B6 | Human | MP | II | Val | Milstein (1969a) |
| Bat | Human | MP | III | | Smithies, Gibson, and Fanning (1972) |
| Bel | Human | MP | I | Val | Milstein (1969b) |
| BJ | Human | MP | I | Leu | Dayhoff (1969) |
| Car(1) | Human | MP | I | | Dayhoff (1969); Milstein, Milstein, and Feinstein (1969) |
| Car(2) | Human | MP | I | | Capra and Kunkel (1970) |
| Cas | Human | MP | II | | Dayhoff (1969) |
| Con | Human | MP | I | | Dayhoff (1969) |
| Cra | Human | MP | I | | Dayhoff (1969) |
| Cum | Human | MP | III | Val | Dayhoff (1969); Hilschmann (1969) |
| Dav[b] | Human | HP | I | | Capra and Kunkel (1970) |
| Dee | Human | MP | I | | Dayhoff (1969) |
| Die | Human | MP | I | | Capra and Kunkel (1970) |
| Dil | Human | MP | II | Val | Smithies, Gibson, and Fanning (1972); Edmundson et al. (1971) |
| Dob | Human | MP | II | | Hood and Terry, unpublished observations. |
| Eu | Human | MP | I | Val | Edelman et al. (1969) |
| Fin[b] | Human | HP | I | | Capra and Kunkel (1970) |

(continued)

**Table A-1. (continued)**

| Protein | Species | Source[a] | Subgroup | Allotype | References |
|---|---|---|---|---|---|
| Fr4 | Human | MP | II | Leu | Milstein (1969a) |
| Gal | Human | MP | III | Leu | Milstein, Jarvis, and Milstein (1969) |
| Gra | Human | MP | II | | Dayhoff (1969) |
| Hac | Human | MP | II | Val | Terry, Hood, and Steinberg (1969); Waterfield, Hood, and Grant, unpublished observations |
| Hau | Human | MP | I | Val | Watanabe and Hilschmann (1970) |
| HBJ 1 | Human | MP | I | | Dayhoff (1969) |
| HBJ 4 | Human | MP | I | Leu | Sox, Stanton, and Hood, unpublished observation; Hood, Grant, and Sox (1970) |
| HBJ 5 | Human | MP | II | | Dayhoff (1969) |
| HBJ 10 | Human | MP | I | | Dayhoff (1969) |
| How | Human | MP | II | | Kaplan and Metzger (1969) |
| HS 4 | Human | MP | II | | Dayhoff (1969) |
| Joh | Human | MP | I | | Capra and Kunkel (1970) |
| Ker | Human | MP | I | Leu | Dayhoff (1969) |
| Lay | Human | MP | I | | Kaplan and Metzger (1969) |
| LPC 1 | Mouse | MP | | | Hood, Potter, and McKean (1971) |
| Lux | Human | MP | I | | Dayhoff (1969) |
| Man | Human | MP | III | Val | Dayhoff (1969) |
| Mar | Human | MP | I | | Kaplan and Metzger (1969) |
| MBJ 41 | Mouse | MP | I | | Dayhoff (1969) |
| MBJ 70 | Mouse | MP | II | | Dayhoff (1969) |
| Mil (HBJ 3) | Human | MP | III | Val | Dayhoff (1969) |
| Mon | Human | MP | I | | Dayhoff (1969) |
| MOPC 8 | Mouse | MP | VI | | Hood, Potter, and McKean (1971) |
| MOPC 15 | Mouse | MP | VI | | Hood, Potter and McKean (1971) |
| MOPC 23 | Mouse | MP | I | | Hood, Potter, and McKean (1971) |

(continued)

**Table A-1. (continued)**

| Protein | Species | Source[a] | Subgroup | Allotype | References |
|---|---|---|---|---|---|
| MOPC 31C | Mouse | MP | I | | Hood, Potter, and McKean (1971) |
| MOPC 37 | Mouse | MP | IV | | Hood, Potter, and McKean (1971) |
| MOPC 46 | Mouse | MP | | | Dayhoff (1969) |
| MOPC 47 | Mouse | MP | IV | | Hood, Potter, and McKean (1971) |
| MOPC 63 | Mouse | MP | II | | Hood, Potter, and McKean (1971) |
| MOPC 149 | Mouse | MP | I | | Hood, Potter, and McKean (1971); Dayhoff (1969) |
| MOPC 167 | Mouse | MP | | | Hood, Potter, and McKean (1971) |
| MOPC 173 | Mouse | MP | I | | Hood, Potter, and McKean (1971) |
| MOPC 265 | Mouse | MP | VII | | Hood, Potter, and McKean (1971) |
| MOPC 321 | Mouse | MP | II | | Hood, Potter, and McKean (1971) |
| MOPC 384 | Mouse | MP | III | | Hood, Potter, and McKean (1971) |
| MOPC 467 | Mouse | MP | IV | | Hood, Potter, and McKean (1971) |
| MOPC 603 | Mouse | MP | III | | Hood, Potter, and McKean (1971) |
| MOPC 674 | Mouse | MP | V | | Hood, Potter, and McKean (1971) |
| MOPC 773 | Mouse | MP | VII | | Hood, Potter, and McKean (1971) |

(continued)

Table A–1. (continued)

| Protein | Species | Source[a] | Subgroup | Allotype | References |
|---|---|---|---|---|---|
| MOPC 843 | Mouse | MP | V | | Hood, Potter, and McKean (1971) |
| MOPC 870 | Mouse | MP | III | | Hood, Potter, and McKean (1971) |
| Nig | Human | MP | II | | Dayhoff (1969) |
| Ou | Human | MP | I | Val | Kohler et al. (1970b) |
| Pap | Human | MP | I | | Dayhoff (1969) |
| Paul | Human | MP | I | | Smithies, Gibson, and Fanning (1972) |
| Pot | Human | MP | I | | Capra and Kunkel (1970) |
| Rad | Human | MP | II | Val | Milstein (1969a) |
| Rai | Human | MP | III | Leu | Milstein, Jarvis, and Milstein (1969) |
| Roy | Human | MP | I | Leu | Dayhoff (1969) |
| Smi | Human | MP | II | | Dayhoff (1969) |
| Ste | Human | CA | II | Val | Edman and Cooper (1968); Cooper and Steinberg (1970) |
| Tei | Human | MP | I | | Capra and Kunkel (1970) |
| Tew | Human | MP | III | Val | Putnam, F.W., personal communication quoted by Smith, Hood, and Fitch (1971) |
| Ti | Human | MP | II | Val | Dayhoff (1969) |
| Tra | Human | MP | I | | Dayhoff (1969) |
| Wag | Human | MP | I | | Kaplan and Metzger (1969) |
| Win | Human | MP | II | | Dayhoff (1969) |
| 1305 | Rabbit | Antibody | | | Jaton et al. (1970) |
| 2436 | Rabbit | Antibody | | | Hood et al. (1970) |
| 2461 | Rabbit | Antibody | | | Hood et al. (1970) |
| 2690 | Rabbit | Antibody | | | Hood et al. (1970) |
| 2711 | Rabbit | Antibody | | | Hood et al. (1970) |
| | Chicken | Normal | | | Dayhoff (1969) |
| | Duck | Normal | | | Dayhoff (1969) |
| | Guinea pig | Normal | | | Dayhoff (1969) |
| | Pig | Normal | | | Dayhoff (1969); Novotny and Franek (1970) |
| | Rabbit | Normal | | *b*4 | Appella et al. (1970); Frangione, Milstein, and Pink (1969) |

(continued)

**Table A-1. (continued)**

| Protein | Species | Source[a] | Subgroup | Allotype | References |
|---|---|---|---|---|---|
| | Rabbit | Normal | | *b*5 | Appella et al. (1970) |
| | Rabbit | Normal | | *b*6 | Appella et al. (1970) |
| | Rat | Normal | | | Dayhoff (1969) |
| | Turkey | Normal | | | Dayhoff (1969) |

[a]MP = myeloma protein or Bence-Jones protein; antibody = purified antibody of restricted heterogeneity from individual animal; normal = normal heterogeneous immunoglobulin from unimmunized individual or pool of individuals; HP = from patients suffering from hypergammaglobulinemia purpura; CA = cold agglutinin against RBC's found in serum of some patients.

[b]Dav and Fin have identical sequences for positions 1 to 40.

**Table A-2.** Properties and provenances of λ chains.

| Protein | Species | Source[a] | Subgroup | Subclass | References |
|---|---|---|---|---|---|
| Bau | Human | MP | III | arg | Baczko et al. (1970) |
| BJ 98 | Human | MP | II | arg | Dayhoff (1969) |
| Bo | Human | MP | IV | arg | Dayhoff (1969) |
| Bu | Human | MP | | | Dayhoff (1969) |
| Daw | Human | MP | | | Dayhoff (1969) |
| Fr | Human | MP | | | Dayhoff (1969) |
| H 2020 | Mouse | MP | | | Weigert et al. (1970) |
| H 2061 | Mouse | MP | | | Weigert et al. (1970) |
| Ha | Human | MP | II | arg | Dayhoff (1969) |
| HBJ 2 | Human | MP | IV | lys | Dayhoff (1969); Hood and Ein (1968b) |
| HBJ 7 | Human | MP | II | arg | Dayhoff (1969); Hood and Ein (1968b) |
| HBJ 8 | Human | MP | I | lys | Dayhoff (1969); Hood and Ein (1968b) |
| HBJ 11 | Human | MP | II | arg | Dayhoff (1969); Hood and Ein (1968b) |
| HBJ 15 | Human | MP | I | lys | Dayhoff (1969); Hood and Ein (1968b) |
| HOPC 1 | Mouse | MP | | | Weigert et al. (1970) |
| Hul | Human | MP | | arg | Edmundson et al. (1968) |
| HS 5 | Human | MP | | lys | Dayhoff (1969) |

(continued)

**Table A-2. (continued)**

| Protein | Species | Source[a] | Subgroup | Subclass | References |
|---|---|---|---|---|---|
| HS 68 | Human | MP | I | arg | Hood and Ein (1968b) |
| HS 70 | Human | MP | I | arg | Hood and Ein (1968b) |
| HS 77 | Human | MP | I | arg | Hood and Ein (1968b) |
| HS 78 | Human | MP | II | arg | Hood and Ein (1968b) |
| HS 86 | Human | MP | I | lys | Hood and Ein (1968b) |
| HS 92 | Human | MP | II | lys | Dayhoff (1969) |
| HS 94 | Human | MP | II | arg | Hood and Ein (1968b) |
| J 558 | Mouse | MP | | | Weigert et al. (1970) |
| J 698 | Mouse | MP | | | Weigert et al. (1970) |
| Kern | Human | MP | III | arg | Dayhoff (1969); Hilschmann, personal communication (revisions) |
| Koh | Human | MP | II | | Kaplan and Metzger (1969) |
| Lb | Human | MP | | | Dayhoff (1969) |
| Li | Human | MP | | | Dayhoff (1969) |
| Lw | Human | MP | | | Dayhoff (1969) |
| Mil | Human | MP | | | Dayhoff (1969) |
| MOPC 104 | Mouse | MP | | | Appella (1971): Weigert et al. (1970) |
| MOPC 315 | Mouse | MP | | | Goetzl and Metzger (1970) |
| Mz | Human | MP | | arg | Dayhoff (1969) |
| New | Human | MP | II | arg | Dayhoff (1969) |
| RPC 20 | Mouse | MP | | | Appella (1971); Weigert et al. (1970) |
| S 176 | Mouse | MP | | | Weigert et al. (1970) |
| S 178 | Mouse | MP | | | Weigert et al. (1970) |
| Sch | Human | MP | | | Dayhoff (1969) |
| Sh | Human | MP | V | arg | Dayhoff (1969) |
| Vil | Human | MP | I | arg | Ponstingl and Hilschmann (1969) |
| Vin | Human | MP | III | lys | Dayhoff (1969); Hess and Hilschmann (1970) |
| We | Human | MP | | | Dayhoff (1969) |
| Wil | Human | MP | | | Dayhoff (1969) |
| X | Human | MP | III | arg | Dayhoff (1969) |

(continued)

**Table A–2. (continued)**

| Protein | Species | Source[a] | Subgroup | Subclass | References |
|---|---|---|---|---|---|
| XP 8 | Mouse | MP | | | Weigert et al. (1970) |
| 111 | Human | MP | III | lys | Langer, Steinmetz-Kayne, and Hilschmann (1968) |
| | Baboon | Normal | | | Hood, Grant, and Sox (1970) |
| | Chicken | Normal | | | Hood, Grant, and Sox (1970); Kubo, Rosenblum, and Benedict (1970) |
| | Chimp | Normal | | | Hood, Grant, and Sox, (1970) |
| | Cow | Normal | | | Hood, Grant, and Sox (1970); Beale and Squires (1970) |
| | Dog | Normal | | | Hood, Grant, and Sox (1970); Dayhoff (1969) |
| | Duck | Normal | | | Hood, Grant, and Sox (1970) |
| | Guinea pig | Normal | | | Hood, Grant, and Sox (1970); Dayhoff (1969) |
| | Horse | Normal | | | Dayhoff (1969) |
| | Monkey | Normal | | | Hood, Grant, and Sox (1970) |
| | Pig | Normal | | | Dayhoff (1969); Franek, Keil, and Sorm (1970); Franek and Novotny (1969); Novotny, Franek, and Sorm (1970); Hood, Grant, and Sox (1970) |
| | Rabbit | Normal | | | Dayhoff (1969) |
| | Sheep | Normal | | | Hood, Grant, and Sox (1970) |
| | Turkey | Normal | | | Hood, Grant, and Sox (1970) |

[a]MP = myeloma protein or Bence-Jones protein; normal = normal heterogeneous immunoglobulin from unimmunized animals.

**Table A–3.** Properties and provenances of H chains.

| Protein | Species | Source* | Subgroup | Class or subclass | Allotype | References |
|---|---|---|---|---|---|---|
| Ale | Human | MP | | $\mu$ | | Pink and Milstein (1967) |
| Bru | Human | MP | | $\gamma3$ | | Frangione, Milstein, and Pink (1969) |
| Ca | Human | MP | I | $\gamma1$ | | Pitcher and Konigsberg (1970) |
| Car | Human | MP | | $\gamma1$ | *Gm*1,17 | Dayhoff (1969) |
| Cor | Human | MP | II | $\gamma1$ | *Gm*1,17 | Press and Hogg (1970) |
| Cra | Human | MP | | $\gamma1$ | *Gm*1 | Dayhoff (1969); Milstein (1969b) |
| Daw | Human | MP | II | $\gamma1$ | *Gm*1,17 | Press and Hogg (1970); Dayhoff (1969) |
| Dee | Human | MP | I | $\gamma1$ | *Gm*1,17 | Dayhoff (1969) |
| Di | Human | MP | I | $\mu$ | | Kohler et al. (1970a) |
| Eik | Human | MP | III | $\gamma1$ | | Capra, personal communication |
| Er | Human | MP | | $\delta$ | | Perry and Milstein (1970) |
| Eu | Human | MP | I | $\gamma1$ | *Gm*4,22 | Edelman et al. (1969) |
| Fie | Human | MP | | $\gamma1$ | | Milstein, Frangione, and Pink (1967) |
| For | Human | MP | III | $\alpha1$ | | Wang et al. (1970a) |
| Ger | Human | MP | | $\gamma4$ | | Prahl (1967) |
| Ha | Human | MP | III | $\alpha1$ | | Kohler et al. (1970a) |
| He | Human | MP | II | $\gamma1$ | *Gm*4,22 | Cunningham et al. (1969) |
| Hin | Human | HCD | III | $\gamma$ | | Terry and Ohms (1970) |
| Kup | Human | MP | | $\gamma3$ | | Frangione and Milstein (1969); Frangione, Milstein, and Pink (1969) |
| Mar | Human | MP | | $\gamma3$ | *Gm*11 | Prahl (1967) |
| MOPC 21 | Mouse | MP | | $\gamma1$ | $Ig\text{-}4^{fast}$ | Svasti and Milstein (1970) |
| MOPC 141 | Mouse | MP | | $\gamma2b$ | $Ig\text{-}3^{a}$ | de Preval, Pink, and Milstein (1970) |
| MOPC 173 | Mouse | MP | | $\gamma2a$ | $Ig\text{-}1^{a}$ | Bourgois and Fougereau (1970 and in preparation); de Preval, Pink, and Milstein (1970) |
| MP 5563 | Mouse | MP | | $\gamma2a$ | $Ig\text{-}1^{a}$ | de Preval, Pink, and Milstein (1970) |
| Na | Human | MP | III | $\mu$ | | Kohler et al. (1970a) |
| Nie | Human | MP | III | $\gamma1$ | *Gm*1, 17 | Ponstingl et al. (1970) |
| Ou | Human | MP | II | $\mu$ | | Dayhoff (1969); Kohler et al. (1970b); Paul et al. (1971); Shimizu et al. (1971) |

| | | | | | | |
|---|---|---|---|---|---|---|
| Sa | Human | MP | | $\gamma 2$ | | Frangione, Milstein, and Pink (1969) |
| She | Human | MP | | $\gamma 4$ | | Prahl (1967) |
| Ste | Human | MP | I | $\gamma 1$ | *Gm*1-4+ | Fisher, Palm, and Press (1969) |
| Til | Human | MP[†] | | $\gamma 2,\mu$ | | Wang et al. (1970b) |
| Vil | Human | MP | | $\gamma 3$ | *Gm*21 | Prahl (1967) |
| Vin | Human | MP | III | $\gamma 4$ | | Dayhoff (1969); Pink and Milstein (1969); Pink et al. (1970) |
| Wan | Human | MP | | $\gamma 2$ | | Prahl (1967) |
| Wat | Human | MP | III | $\gamma 2$ | | Wang et al. (1970a) |
| Wo | Human | MP | III | $\mu$ | | Kohler (1970) |
| Zuc | Human | HCD | III | $\gamma 3$ | *Gm*11 | Dayhoff (1969); Frangione and Milstein (1969) |
| Un-named (1) | Human | MP | | $\gamma 1$ | *Gm*1+4- | Thorpe and Deutsch (1969) |
| Un-named (2) | Human | MP | | $\gamma 1$ | *Gm*1+4- | Frangione et al. (1966) |
| | Human | MP | | $\alpha$ | | Abel and Grey (1967) |
| | Human | MP | | $\mu$ | | Abel and Grey (1967) |
| | Cow | Normal | | $\gamma 1,\gamma 2$ | | Dayhoff (1969) |
| | Guinea pig | Normal | | $\gamma 2$ | | Cebra et al. (1970) |
| | Horse | Normal | | $\gamma,\gamma(T)$ | | Dayhoff (1969) |
| | Mouse | MP | | $\alpha$ | | Abel and Grey (1967) |
| | Baboon and rhesus monkey | Normal | | $\gamma$ | | Wang and Fudenberg (1969) |
| | Rabbit | Normal | | $\gamma$ | *a*1,*a*3 | Wilkinson (1969a) |
| | Rabbit | Normal | | $\gamma$ | *a*11,*a*12, *a*14,*a*15 | Dayhoff (1969); Fruchter et al. (1970); O'Donnell, Frangione, and Porter (1970); Prahl, Mandy, and Todd (1969); Appella et al. (1971) |

*MP = myeloma protein; normal = normal heterogeneous immunoglobulin from unimmunized animal; HCD = "heavy chain disease" proteins, carrying large deletions in H chains.

[†]Two cell lines in the same patient.

# Appendix B

## Immunoglobulin Allotypes

Allelic differences in immunoglobulin structures are called *allotypes.* Most of these allotypes are detected by specific antisera, although electrophoretic mobility and direct sequence or peptide mapping analysis have also been used to identify them. The details of the detection systems need not concern us here and can be found in the cited references.

In very few cases has it been firmly established that these serologically assayed allelic differences result from sequence differences in the structural genes to which they are formally assigned. In few cases, for example, has it been excluded that they are due to carbohydrate moieties, or to other associated polypeptides. In fact, one report (Lobb, Curtain, and Kidson, 1967) suggests that the actual genes involved are "control" genes which determine which of several allotypic specificities, all present in all individuals, will actually be expressed in any particular individual. In this book I have taken the allotypes at their face value, however, for to do so is to build up a consistent picture of the genetic organization of immunoglobulin structural genes, which fits neatly into the independent "family" organization inferred in Chapter 9 from entirely different considerations. The only significant conclusions that have been drawn from the allotypes (apart from the considerations of Chapter 7) are first, their organization into gene families and second, the predominantly intrachromosomal association of V and C genes (Chapter 9).

All the allotypes discussed herein behave as if they were determined by single mendelian loci. However, I have scrupulously avoided concluding that the polypeptide chains carrying a particular allotype derive from a single germline gene. This conclusion is neither justified nor necessary. Nevertheless, since it is simplest to assume that each C gene occurs once per haploid genome, I shall speak in this appendix as if these genes are singular.

In compiling this review of allotypes, I have been aided immeasurably by a recent definitive treatise on the human allotypes by Grubb (1970), and a review of the mouse allotypes by Herzenberg, McDevitt, and Herzenberg (1968). Recent work on allotypes in wild mice (Lieberman and Potter, 1969) has not yet been analyzed unambiguously at the molecular level and therefore is excluded.

Each allotype is assigned to a structural gene locus on the basis of its molecular location. Most of the allotypes are found in the Fc domain of one and only one of the known immunoglobulin classes or subclasses: these are accordingly assigned to the Fc segment of the corresponding $C_H$

structural gene. The Fab determinants are assigned to either $C_L$, $V_H$, or $C_H1$ on the basis of their location in isolated L or H chains, and/or their presence in different immunoglobulin classes (which differ in $C_H$ but not in $C_L$ or $V_H$). Except for the two allotypes which must be assigned to V regions (see Chapter 7), determinants have been assigned to C rather than V regions where a choice was possible. All these assignments are consistent with the sequence differences which have been correlated with some of the allotypes.

Appendix Tables B–1 and B–2 list the allotypes of man and rabbit respectively (although some have been omitted), their assigned loci, positions at which correlated sequence variation has been found, their linkage relationships, and references. The allotypes of the mouse can be found in the Herzenberg review cited above.

The presence of one allotypic specificity on a molecule does not necessarily exclude the presence of others. Rather, for each locus alternative *patterns* of specificities are found. In the mouse, these patterns (listed in Appendix Table B–3) are defined by the determinants present in immunoglobulins of inbred strains; in humans, they are defined by examining the patterns of determinants present on individual myeloma proteins (see Table B–4). In the rabbit, the allotypes for the most part have been analyzed as mutually exclusive pairs (or triplets) of markers, as they are listed in Table B–2. Some of the allotypes in this species may in fact be families of different specificities, associated together.

In each of the three species the allotypes can be arranged into genetic linkage groups, as determined by genetic crosses, pedigree analyses, or population studies. The linkage groups so inferred were shown in the genetic map in Fig. 9–2. From it the important generalization emerges that structural genes—either V or C—in the same immunoglobulin gene family are closely linked to one another. References for the linkage relationships for man and rabbit are found in Appendix Tables B–1 and B–2 respectively. In the mouse, over two thousand backcross progeny have been assayed for recombination among the four $C_H$ genes for which allotypes are known, and none have been found (Herzenberg, McDevitt, and Herzenberg, 1968).

In man, part of the evidence for linkage of the genes in the *Gm* linkage group (controlling H chains) is the existence of "gene complexes" in human populations. A complex is a group of allotypes which are far more commonly found in association with one another in the same individuals than expected by chance if they assort independently; consequently, they must be closely linked. The major gene complexes in human populations are shown in Appendix Table B–4. Similarly, among 68 inbred mouse strains, each of which represents a haploid genotype, there are only eight distinct patterns of H-chain allotypes in the *Ig* linkage group, implying that in this species, too, H-chain genes are associated into gene complexes (Appendix Table B–3).

The extremely low frequency of recombination between linked

**Table B-1.** Allotypes of human immunoglobulins.

| Allotype | Former names | Molecular location | Assigned gene | Positions of correlated sequence differences | Linkage relations | Reference (see below) |
|---|---|---|---|---|---|---|
| *Gm* 1 | *Gm* a | Fc of $IgG_1$ | $C_{\gamma 1}(C_H3)$ | 356,358 | | 1 |
| *Gm* 2 | *Gm* x | Fc of $IgG_1$ | $C_{\gamma 1}(Fc)$ | 356,358 | Linked to *Gm* 1 | 1 |
| *Gm* 3 | *Gm* $b^w, b^2$ | Fd of $IgG_1$ | $C_{\gamma 1}(C_H1)$ | 214 | Linked to *Gm* 5, antithetical to *Gm* 1 | 1 |
| *Gm* 4 | *Gm* f | Fd of $IgG_1$ | $C_{\gamma 1}(C_H1)$ | 214 | Antithetical to *Gm* 1, linked to *Gm* 5; identical to *Gm* 3 (?) | 1 |
| *Gm* 5 | *Gm* b, *Gm* b′ | Fc of $IgG_3$ | $C_{\gamma 3}(C_H3)$ | 336 | Antithetical to *Gm* 1 | 1 |
| *Gm* 6 | *Gm* c, *Gm*-like | $IgG_3$ | $C_{\gamma 3}$ | | Linked to *Gm* 5 | 1 |
| *Gm* 8 | *Gm* e | $IgG_1$ and $IgG_2$ | ? | | Linked to *Gm* 1,2,5 (may be "non-a") | 1 |
| *Gm* 11 | *Gm* $b^\beta, b^0$ | $IgG_3$ | $C_{\gamma 3}$ | | Linked to *Gm* 5 | 1 |
| *Gm* 12 | *Gm* $b^\gamma$ | $IgG_3$ | $C_{\gamma 3}$ | | Linked (identical?) to *Gm* 5 | 1 |
| *Gm* 13 | *Gm* $b^3$ | $IgG_3$ | $C_{\gamma 3}$ | | Linked to *Gm* 5 | 1 |
| *Gm* 14 | *Gm* $b^4$ | $IgG_3$ | $C_{\gamma 3}$ | | Linked to *Gm* 5; antithetical to *Gm* 6 in Negroes | 1 |
| *Gm* 16 | *Gm* t | $IgG_3$ | $C_{\gamma 3}$ | | Linked to *Gm* 13 | 1 |
| *Gm* 17 | *Gm* z | Fd of $IgG_1$ | $C_{\gamma 1}(C_H1)$ | 214 | Linked to *Gm* 1 | 1 |
| *Gm* 18 | Ig $Ro_2$ | Fc of $IgG_1$ | $C_{\gamma 1}(Fc)$ | | Linked to *Gm* 1, 2 | 1 |
| *Gm* 21 | *Gm* g | $IgG_3$ | $C_{\gamma 3}$ | 336 | Antithetical to *Gm* 11, linked to *Gm* 1 | 1 |
| *Gm* 22 | *Gm* y | Fc of $IgG_1$ | $C_{\gamma 1}(Fc)$ | | Linked to *Gm* 4 | 1 |
| *Gm* 23 | *Gm* n | Fc of $IgG_2$ | $C_{\gamma 2}(Fc)$ | | Linked to *Gm* 4 | 1 |
| "1–3–4" | | Fc of $IgG_1$, | $C_{\gamma 4}$ | | Presence on $IgG_4$ | 2 |

| | | | | | |
|---|---|---|---|---|---|
| | | $IgG_3$, some $IgG_4$ | | | antithetical to "2–4" and *Gm* 23 | |
| "2–4" | | Fc of $IgG_2$, some $IgG_4$ | $C_{\gamma 4}$ | | Presence on $IgG_4$ antithetical to "1–3–4," linked to *Gm* 23 | 2 |
| *Am* 1 | *Am* 2 | α chain of $IgA_2$ | $C_{\alpha 2}$ | Interchain disulfide difference; see Table 2–5 | Linked to *Gm* | 1 |
| *Inv* 1 | *Inv* 1 | | $C_{\kappa}$ | 191 | Linked to *Inv* 2 | 1 |
| *Inv* 2 | *Inv* +, *Inv* a | | $C_{\kappa}$ | | Linked to *Inv* 1, unlinked to *Gm* | 1 |
| *Inv* 3 | *Inv* b | | $C_{\kappa}$ | 191 | Antithetical to *Inv* 1 | 1 |


REFERENCES:

1. Grubb (1970).
2. Kunkel et al. (1970).

**Table B–2.** Allotypes of rabbit immunoglobulins.

| Allotypes | Molecular location | Assigned gene | Correlated sequence differences | Linkage relations | References (see below) |
|---|---|---|---|---|---|
| *a*1,*a*2,*a*3 | Fd of $\gamma,\mu,\alpha,\epsilon$ | $V_H$ | See Chapter 7 | | See Chapter 7 |
| *b*4,*b*5,*b*6,*b*9 | $\kappa$ | $C_\kappa$ | Positions 210,212, and 213 (Table B–5) | Unlinked to *a*1-*a*3 | 7–12 |
| | | | Amino acid composition differences | | 1,2 |
| | | | Peptide map differences | | 3,4 |
| | | | Sequence differences | | 5,6 |
| *c*7,*c*21 | $\lambda$ | $C_\lambda$[a] | | Unlinked to *a* or *b* | 13–16 |
| *a*11,*a*12 | Hinge of $\gamma$ | $C_\gamma$(hinge) | Position 225 (Table B–7) | Linked to *a*1,*a*2,*a*3 | 17–20 |
| *a*14,*a*15 | Fc of $\gamma$ | $C_\gamma(C_H2)$ | Position 309 (Table B–7) | Linked to *a*1,*a*2,*a*3 | 21,22 |
| *a*8,*a*10 | Fc of $\gamma$ | $C_\gamma$(Fc) | | Linked to *a*1,*a*2,*a*3 | 23,26 |
| *f*1–*f*5 | IgA in serum and colostrum | Probably secretory piece | | Unlinked to *a* or *b* | 14,27,28 |
| "c–1" | Fc of IgA in serum and colostrum | $C_\alpha$ | | | 29 |
| *Ms*1,*Ms*2 | IgM | $C_\mu$(?) | | | 30,31 |
| "$\mu^1,\mu^2$" | IgM | $C_\mu$ | | | Alluded to in 22 |

[a]Usually *c*7 and *c*21 are allelic and on separate molecules; however, Gilman-Sachs et al. (1969) report one animal in which *c*7 and *c*21 are not allelic but linked.

REFERENCES


1. Reisfeld, Dray, and Nisonoff (1965).
2. Reisfeld and Inman (1968).
3. Frangione, Franklin, and Kelus (1968).
4. Small, Reisfeld, and Dray (1965).

17. Mandy and Todd (1969).
18. Prahl et al. (1970).
19. Prahl, Mandy, and Todd (1969).
20. Zullo, Todd, and Mandy (1968).

5. Appella, Rejnek, and Reisfeld (1969).
6. Frangione (1970).
7. Dubiski and Muller (1967).
8. Oudin (1960a).
9. Oudin (1960b).
10. Sell and Hughes (1968).
11. Stemke (1964).
12. Wilheim and Lamm (1966).
13. Chersi et al. (1970).
14. Gilman-Sachs et al. (1969).
15. Mage, Young, and Reisfeld (1968).
16. Vice, Hunt, and Dray (1970).
21. Dubiski (1969).
22. Mage, Young-Cooper, and Alexander (1971).
23. Hamers and Hamers-Casterman (1965).
24. Hamers and Hamers-Casterman (1967).
25. Hamers, Hamers-Casterman, and Kelus (1965).
26. Hamers, Hamers-Casterman, and Lagnaux (1966).
27. Conway, Dray, and Lichter (1969a).
28. Conway, Dray, and Lichter (1969b).
29. Masuda, Kuribayashi, and Hanaoka (1969).
30. Kelus and Gell (1965).
31. Sell (1966).

**Table B–3.** Alternative patterns of allotypes in mouse immunoglobulins, and the restricted number of gene complexes found among 68 inbred strains. No recombinants have been detected in over 2,000 backcross progeny. (Taken from Herzenberg, McDevitt, and Herzenberg, 1968.) Each pattern is characterized by the presence of several individual determinants, and the absence of others.

| Gene complex | Number of strains | Alternative patterns of genes– | | | |
|---|---|---|---|---|---|
| | | $C_{\gamma}2a$ | $C_{\alpha}$ | $C_{\gamma}2b$ | $C_{\gamma}1$ |
| a | 25 | Ig-$1^{a}$ | Ig-$2^{a}$ | Ig-$3^{a}$ | Ig-$4^{fast}$ |
| b | 18 | Ig-$1^{b}$ | Ig-$2^{b}$ | Ig-$3^{b}$ | Ig-$4^{slow}$ |
| c | 6 | Ig-$1^{c}$ | Ig-$2^{c}$ | Ig-$3^{a}$ | Ig-$4^{fast}$ |
| d | 2 | Ig-$1^{d}$ | Ig-$2^{d}$ | Ig-$3^{d}$ | Ig-$4^{fast}$ |
| e | 4 | Ig-$1^{e}$ | Ig-$2^{d}$ | Ig-$3^{e}$ | Ig-$4^{fast}$ |
| f | 4 | Ig-$1^{f}$ | Ig-$2^{f}$ | Ig-$3^{f}$ | Ig-$4^{fast}$ |
| g | 4 | Ig-$1^{g}$ | Ig-$2^{c}$ | Ig-$3^{g}$ | Ig-$4^{fast}$ |
| h | 5 | Ig-$1^{h}$ | Ig-$2^{a}$ | Ig-$3^{a}$ | Ig-$4^{fast}$ |

immunoglobulin loci makes it impossible to map their order by measuring recombination rates. However, the three-factor cross method, based on the assumption that double crossovers are much rarer than single crossovers in the same region, can theoretically allow the order of tightly linked genes to be ascertained with only a few recombinants. For example, the single crossover observed in the rabbit *a* linkage group suggests that the $C_{\mu}$ gene is between the $V_H$ gene(s) and the $C_{\gamma}$ gene, as shown in Appendix Fig. B–1. In humans, unfortunately, illegitimacy of such rare recombinants cannot be excluded with convincing certainty; for the probability of non-paternity would have to be much less than the severest critic's most niggardly guess at a recombination rate. Nevertheless the gene order $[C_{\gamma 2}.C_{\gamma 3}.C_{\gamma 1}]$ suggested by the three reported possible human recombinants (Natvig, Kunkel, and Litwin, 1967) is consistent with the gene order $[C_{\gamma 2}.(C_{\gamma 3},C_{\gamma 1})]$ implied by the "Lepore"-type hybrid immunoglobulin shown in Appendix Fig. B–2. A hybrid myeloma protein has also been described in the mouse (Warner, Herzenberg, and Goldstein, 1966).

This ordering of the $C_H$ genes on the basis of rare recombinants, however, assumes that the genes are in linear array; as discussed in Chapter 9, they might in fact form a lateral array. If so, we may eventually find three-factor crosses whose results are inconsistent with a single tandem array of genes.

The weight of the evidence, then, suggests that the human and mouse $C_H$ genes are arranged in linked arrays on their chromosomes. This configuration is expected if $C_H$ genes arose by gene duplications resulting from either unequal crossing over or the formation of lateral branches as proposed in Smithies' hypothesis.

**Table B-4.** Alternative patterns of allotypes in human immunoglobulins and some common gene complexes in human populations. (Taken from Grubb, 1970; Kunkel et al., 1970; Natvig, Kunkel, and Litwin, 1967.)

| | Alternative patterns of *Gm* determinants | | | | | Alternative patterns of *Inv* determinants | |
|---|---|---|---|---|---|---|---|
| | $C_{\gamma 4}$ | $C_{\gamma 2}$ | $C_{\gamma 3}$ | $C_{\gamma 1}$ | | $C_{\kappa}$ | |
| | Fc | Fc | Fc | $C_H1$ | Fc | | |
| Caucasian | "1–3–4" | 23– | 21 | 17 | 1 | 1 | present with different |
| *Gm* gene | "1–3–4" | 23– | 21 | 17 | 1,2 | 1,2 | frequencies in |
| complexes | "1–3–4" | 23– | 5,11,13,14 | 4 | 22 | 3 | most populations |
| | "2–4" | 23+ | 5,11,13,14 | 4 | 22 | | |
| Negro | | 23– | 5,11,13,14 | 17 | 1 | | |
| *Gm* gene | Not | 23– | 5,6,11 | 17 | 1 | | |
| complexes | determined | 23– | 5,6,11,14 | 17 | 1 | | |
| | | 23– | 11,13 | 17 | 1 | | |
| Mongoloid | Not | 23– | 11,13 | 17 | 1 | | |
| *Gm* gene | determined | 23+ | 5,11,13,14 | 4 | 1 | | |
| complexes | | | | | | | |

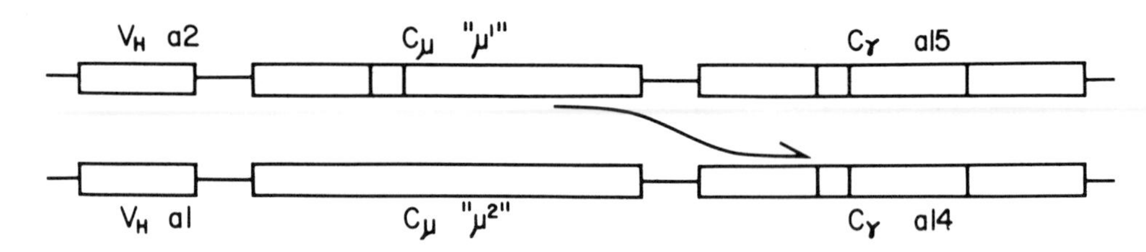


**Figure B–1.** A probable crossover in the rabbit *a* linkage group (Mage, Young-Cooper, and Alexander, 1971).

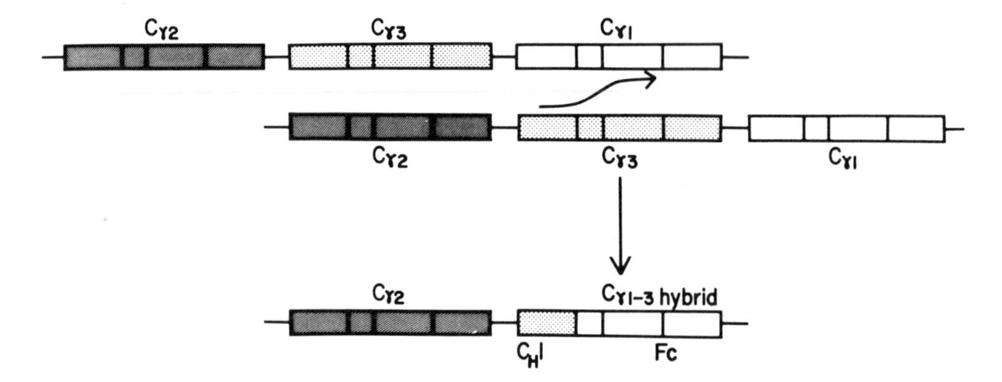


**Figure B–2.** A human family with a hybrid $C_\gamma$ region (Kunkel, Natvig, and Joslin, 1969; Steinberg, Muir, and McIntire, 1968). The hybrid molecule had subclass-specific determinants for $IgG_3$ in the Fab region and for $IgG_1$ in the Fc region. If the polarities of the genes were reversed, the hybrid should have $IgG_1$ Fab determinants and $IgG_3$ Fc determinants. Since $IgG_2$ was present in normal amounts, the $C_{\gamma 2}$ gene probably does not lie between the $C_{\gamma 3}$ and $C_{\gamma 1}$ genes.

All the sequence variation which has been observed within the known C-region classes or subclasses is shown in Appendix Tables B–5 to B–7. Most of the variation correlates with known allelic differences, as is evident from the tables, and permits specific sequence variants to be assigned to alternative alleles as shown in Appendix Table B–8.

In Appendix Tables B–5 to B–7 all variants which do not correlate with known allotypic divergences are printed in boldface; they are very few, despite the fact that many independent C-region sequences from several human and mouse classes and subclasses have been determined (Tables 2-2B, 2–3B, and 2–4B). These unexplained variants are comparable to variants in the human hemoglobin $\alpha$ and $\beta$ chains which have been studied for years (see Dayhoff, 1969). Like them, they are probably a result of recent allelic mutations or gene duplications which have not been detected by conventional methods (either because they have not led to detectable

**Table B-5.** Variation in $C_{\kappa}$ regions. Unexplained variant is in boldface.

| Species | Allotype | | Protein | Residue at position– | | | | |
|---|---|---|---|---|---|---|---|---|
| | | | | 210 | | 212 | | 213 |
| Rabbit | *b*4 | | Normal | D | | G | | N |
| | *b*5 | | Normal | S | | K | | N |
| | *b*6 | | Normal | S | | K | | S |
| Human | *Inv* 1/2[a] | *Inv* 3 | | | 122 | | 191 | |
| | + | – | Rai | | | | L | |
| | + | – | Ja[b] | | | | L | |
| | + | – | Gal | | | | L | |
| | + | – | Fr4 | | | | L | |
| | | | HBJ4 | | B | | L | |
| | + | – | Roy | | D | | L | |
| | + | – | Ker | | D | | L | |
| | + | – | BJ | | D | | L | |
| | | | Ag | | **N** | | V | |
| | – | + | Cum | | D | | V | |
| | | + | Rad | | D | | V | |
| | – | + | Ti | | D | | V | |
| | – | | Eu | | D | | V | |
| | | | Tew | | B | | V | |
| | | | Mil | | B | | V | |
| | | | Dil | | B | | V | |
| | | | Hau | | B | | V | |
| | | | Ou | | B | | V | |
| | | | Au | | B | | V | |
| | – | + | Man | | | | V | |
| | – | + | Hac | | | | V | |
| | | | B6 | | | | V | |
| | – | + | Ste | | | | V | |
| | | | Bel | | | | V | |
| | – | | BJ 26[c] | | | | V | |

[a]Because of considerable confusion about whether earlier reports were dealing with closely related *Inv* 1 or *Inv* 2 markers (see Grubb, 1970), any protein reported as 1+, 2+, 1+2+, or a+ is listed here as ½+. For further discussion of the correlation of *Inv* markers with the residue at 191, see Baglioni et al. (1966); Milstein (1966a, b, and c); Milstein, Frangione, and Pink (1967); and Terry, Hood, and Steinberg (1969). There appear to be no sequence data on κ chains corresponding to the rare genotype 1+2–3–; the sequence differences responsible for the presence or absence of *Inv* 2 are therefore not known.

[b]Reported in Terry, Hood, and Steinberg (1969).

[c]Reported in Baglioni et al. (1966).

**Table B-6.** Variation in $C_\lambda$ regions. Unexplained variants are in boldface.

| Species | Subclass | Protein | Residue at position– | | | |
|---|---|---|---|---|---|---|
| | | | 148 | 157 | 176 | 194 |
| Man | $C_\lambda$Arg | HBJ 11 | | | | R |
| | | HBJ 7 | | | | R |
| | | HS 94 | | | | R |
| | | HS 78 | | | | R |
| | | HS 68 | | | | R |
| | | HS 77 | | | | R |
| | | HS 70 | | | | R |
| | | Hul | | S | K | R |
| | | Ha | A | S | K | R |
| | | New | A | S | K | R |
| | | Bo | A | S | K | R |
| | | Vil | A | S | K | R |
| | | Sh | A | S | K | R |
| | | Bau | A | S | K | R |
| | | X | A | S | K | R |
| | | Kern | A | **G**[a] | K | R |
| | | Mz | **V** | S | **N** | R |
| | $C_\lambda$? | 119[a] | | **G**[a] | | |
| | | Mil | A | | | |
| | | Fr | A | | | |
| | | Sch | A | | | |
| | | Bu | A | | K | |
| | | Wil | A | | K | |
| | | Lw | A | | K | |
| | | Li | A | | K | |
| | | We | A | | K | |
| | | Lb | A | | K | |
| | $C_\lambda$Lys | HS 5 | A | S | K | K |
| | | HS 92 | A | S | K | K |
| | | Vin | | | | K |
| | | HBJ 2 | | | | K |
| | | HBJ 8 | | | | K |
| | | HS 86 | | | | K |
| | | HBJ 15 | | | | K |
| | | 111 | | | | K |
| | | | | | 217 | |
| Mouse | | MOPC 104 | | | S | |
| | | RPC 20 | | | S | |
| | | MOPC 315 | | | **L** | |

[a]Protein 119 is reported by Hess and Hilschmann (1970) who, along with Gibson, Levanon, and Smithies (1971), suggest that the Ser-Gly interchange at position 157 is caused by a duplication of the $C_{\lambda arg}$ gene.

**Table B–7.** Variation in $C_H$ regions of a single class or subclass. Unexplained variants are in boldface.

| Species | Class or subclass | Allotype* | Protein | Residue at position– | | |
|---|---|---|---|---|---|---|
| | | | | 136 | | |
| Mouse | γ2a | *Ig*-1$^{a}$ | MOPC 173 | **T** | | |
| | | *Ig*-1$^{a}$ | MP 5563 | **S** | | |
| | | | | 225 | 309 | |
| Rabbit | γ | *a*11 | Normal | M | | |
| | | *a*12 | Normal | T | | |
| | | *a*14 | Normal | | T | |
| | | *a*15 | Normal | | A | |
| | | | | 214 | 356 | 358 |
| Human | γ1 | *Gm* 4, 22 | Cra | | E | M |
| | | *Gm* 4, 22 | Eu | R | E | M |
| | | *Gm* 4, 22 | He | R | | |
| | | *Gm* 1, 17 | Unnamed (1) | | D | L |
| | | *Gm* 1, 17 | Unnamed (2) | | D | L |
| | | *Gm* 1, 17 | Car | K | | |
| | | *Gm* 1, 17 | Cor | K | | |
| | | *Gm* 1, 17 | Dee | K | | |
| | | *Gm* 1, 17 | Daw | K | | |
| | | | | 436 | | |
| | γ3 | *Gm* 11 | Zuc | F | | |
| | | *Gm* 11 | Mar | F | | |
| | | *Gm* 21 | Vil | Y | | |

*Authors seldom report exactly which allotypes have been tested. The human proteins here have been assigned one of the alternative patterns in Appendix Table B–4 consistent with the specificities reported. The two mouse proteins were assigned the pattern characteristic of their strains (BALB/c for MOPC 173, C3H for MP5563); since these were the same, their difference might lie in some unknown $C_H1$ polymorphism between the two strains.

**Table B–8.** Amino acid sequence differences which correlate with allotypes.

| Species | Gene | Allele | Residue at position– | | |
|---|---|---|---|---|---|
| | | | 210 | 212 | 213 |
| Rabbit | $C_\kappa$ | *b*4 | D | G | N |
| | | *b*5 | S | K | N |
| | | *b*6 | S | K | S |
| | | | 225 | 309 | |
| Rabbit | $C_\gamma$(hinge) | *a*11 | M | | |
| | | *a*12 | T | | |
| Rabbit | $C_\gamma$(Fc) | *a*14 | | T | |
| | | *a*15 | | A | |
| | | | 136 | | |
| Mouse | $C_\gamma 2a$ | ? | T | | |
| | | ? | S | | |
| | | | 191 | | |
| Man | $C_\kappa$ | *Inv* 1/2 | L | | |
| | | *Inv* 3 | V | | |
| | | | 214 | 356 | 358 |
| Man | $C_{\gamma_1}(C_H1)$ | *Gm* 17 | K | | |
| | | *Gm* 4 | R | | |
| Man | $C_{\gamma_1}(C_H3)$ | *Gm* 1 | | D | L |
| | | *Gm* 22 | | E | M |
| | | | 436 | | |
| Man | $C_{\gamma_3}(C_H3)$ | *Gm* 11 | F | | |
| | | *Gm* 21 | Y | | |

antigenic differences or because they are not present in the population with appreciable frequency). For example, there is evidence that the Ser-Gly interchange at position 157 of human $C_{\lambda arg}$ (see Appendix Table B–6) arises from a recent duplication of the $C_{\lambda arg}$ gene, one of the copies harboring the Gly variant (Gibson, Levanon, and Smithies, 1971; Hess and Hilschmann, 1970).

# Appendix C

## The Ribosomal RNA Genes

A multiplicity of related genes is not a feature peculiar to the immunoglobulins, but a general characteristic of all eucaryotic organisms (Britten and Kohne, 1968). Two other multigene systems in higher organisms have been studied in some detail: the linked genes specifying the 18*s* and 28*s* subunits of ribosomal RNA (the rDNA system) and the genes coding the 5*s* subunit of ribosomal RNA (the 5*s*DNA system).

### The rDNA System

There are about five hundred linked, repeated genetic units per haploid genome in the rDNA system of the frog *Xenopus laevis*, each unit consisting of the DNA complements of the 18*s* and 28*s* RNA subunits plus an untranscribed "spacer" sequence (Wallace and Birnstiel, 1966; Brown and Weber, 1968; Wensink and Brown, 1971). This rDNA comprises the nucleolar organizer, around which the nucleolus forms. By contrast, oocytes contain, in addition to the one expected chromosomal complement of five hundred copies, about a million copies in the form of extrachromosomal DNA circles (Miller and Beatty, 1969; see also references in Brown and Blackler, 1972). A nucleolus forms about each of these extrachromosomal circles; one circle contains many copies of rDNA genes. The as-yet unknown process which results in these extrachromosomal copies is called amplification. The amplified rDNA is not maternally inherited and therefore presumably arises from the chromosomal rDNA (Brown and Blackler, 1972).

All the repeated units of rDNA in one organism (including the spacer) are very similar to one another, if not identical. The transcribed part of the repeated unit is highly conserved in evolution, being very similar from one species to another. The spacer portion, on the other hand, can be quite different, even between species as closely related as the cross-fertile frogs *Xenopus laevis* and *X. mulleri* (Brown, Wensink, and Jordan, 1972).

Buongiorno-Nardelli, Amaldi, and Lava-Sanchez (1972) have recently found that the numbers of rDNA genes per haploid genome in various amphibian species assume values very close to the power series of two—512, 1024, 2048, or 4096. This striking pattern adumbrates an earlier observation of H.-G. Keyl (cited by Callan, 1967). He compared the DNA contents of homologous chromomeres (bands) in the salivary chromosomes of heterozygotes formed from the two flies *Chironomus thummi thummi* and

*Ch. th. piger.* The ratios always conformed to the power series of two—1:2, 4, 8, or 16.

The same power series appears again when the number of extrachromosomal nucleoli (thus circles of extrachromosomal rDNA) per oocyte is determined. Buongiorno-Nardelli and his co-workers, in the paper cited above, observed that the numbers of nucleoli in oocytes of the same amphibians fell into three well-defined groups: about 250, 500, or 1000. In most of the species studied the number of nucleoli was less than the diploid number of chromosomal rDNA genes, in three it was equal, and in none was it greater.

## The 5sDNA System

The repeated unit in the 5sDNA system consists of the complement of the 5*s*RNA subunit plus a spacer (Brown, Wensink, and Jordan, 1971). This unit is repeated about twenty-five thousand times per haploid genome in *Xenopus laevis*, but it has not been demonstrated that all these repeats are clustered at a single chromosomal locus. In human HeLa cells multiple clusters of 5sDNA repeats are located on different chromosomes (Aloni, Hatlen, and Attardi, 1971), while in the fruit fly these genes are all located on a single chromosome (Wimber and Steffensen, 1970). As in the rDNA system, the transcribed parts of the 5sDNA repeats are very similar or identical within one organism (Brown, Wensink, and Jordan, 1971) and from species to species. Williamson (1970) has shown that the sequences from *Xenopus* and man differ at less than 10 percent of the nucleotide positions. Although all the spacer sequences within one species are similar to one another, there is some evidence that in *Xenopus laevis* they are considerably more heterogeneous than the transcribed sequences (about 3 percent average difference between spacers versus nearly no difference between transcribed portions, according to Brown, 1971). Since it has not been shown that the *Xenopus* 5sDNA repeats are all clustered at one locus, a possible interpretation of the apparent heterogeneity of spacer sequences is that all those in one locus are as homogeneous as the transcribed sequences, the observed differences lying between spacers in different loci.

These two systems of repeated genes pose an evolutionary problem similar to that posed by the germline theory of antibody diversity. The repeated genes are highly redundant; indeed, in the rDNA and 5sDNA systems, unlike the antibody system, all the genes presumably perform exactly the same function. It is difficult to imagine that the selective force acting on any one redundant gene is significant. Why then do they not accumulate mutations until so few functional genes remain that the loss of any more would endanger survival?

Several theories have been proposed to account for the evolution of these multigene systems. All of them result in the evolution of an entire multigene family as if it were a single gene.

### Rectification by Amplification

Buongiorno-Nardelli, Amaldi, and Lava-Sanchez (1972) propose a theory for the evolution of rDNA genes which at the same time elegantly rationalizes the striking pattern they found in numbers of rDNA genes and nucleoli. They suggest that each extrachromosomal circle of rDNA in the oocyte arises from a single one of the chromosomal rDNA genes. The chromosomal gene is first excised from the chromosome as a circle, and then is repeatedly duplicated by the "opposing rolling circle" mechanism (see Watson, 1970), in which each round of DNA replication doubles the length of the circle. Thus the number of rDNA genes per extrachromosomal circle would be given by $2^n$, where $n$ is the number of rounds of replication. One such extrachromosomal circle is then reintegrated into each of the chromosomal rDNA sites. Because all the copies at any one site are duplicates of one of the original rDNA genes, this process would have the effect of making all the genes at one site identical at each female meiosis; in an evolutionary perspective, the multiple rDNA genes would appear to evolve as a single gene. This proposal predicts that the number of genes per extrachromosomal circle of rDNA equals the number of genes at each chromosomal rDNA site.

At the time of rDNA amplification, the oocyte is tetraploid. If each rDNA gene gave rise to a nucleolus, there should be four times as many nucleoli as the haploid complement of rDNA genes. In fact, however, there are at most only twice as many nucleoli as haploid genes. The authors accordingly assume that only one of the two parental chromosome-pairs gives rise to extrachromosomal rDNA. To explain the many cases where the nucleoli amount to even less than twice the haploid rDNA genes, they propose the subsidiary hypothesis that in those species there are several chromosomal rDNA sites, only one of which gives rise to extrachromosomal DNA.

This theory has the defect of requiring amplification for rectification. Therefore it cannot apply without some modification to the 5sDNA system, whose genes are not amplified, and it is completely inappropriate for multiple hypothetical V genes since it would eliminate diversity in them. Even some rDNA genes—those on the Y chromosome of the fruit fly—never pass through an oocyte and would not be subject to this mode of rectification.

The most important evidence cited by the authors in favor of their theory is the relationship between the number of rDNA genes and the number of nucleoli. Except that both are powers of two, however, there is in fact no correlation between the two sets of values. The authors' explanation of this discrepancy (that only one of several rDNA sites takes part in amplification) seems to require that the number of rDNA sites be a power of two in order to explain why the number of nucleoli is itself a power of two. In short, these investigators have discovered an important clue to the evolution of multigene systems, but it seems very unlikely that their interpretation of the clue is correct.

## The Unequal Crossing Over Model

Unequal crossover, such as is diagrammed in Fig. 1–3, results in the duplication or deletion of a subset of contiguous genes. We will suppose that this process occurs continuously and randomly in a multigene family, each event involving only a small proportion of the genes, and the total number of genes never varying greatly from the starting number. If this process occurs often enough, essentially all of the surviving genes will be descendants of a single one of the original genes.

A rough estimate of the number of crossover events required to reach this state of affairs can be made by analogy with ordinary population genetics. In a randomly breeding population of $N$ organisms (that is, $2N$ chromosomes), it requires about $4N$ generations for the descendants of one of the original chromosomes to take over the entire population. It remains for us to determine some rough equivalent of one generation in terms of crossover events. One generation in ordinary population genetics is approximately the time for each chromosome that is going to survive to leave two descendants, and we will take as its equivalent the number of crossover events required on the average to duplicate every gene in the family which survives. For the sake of illustration, let us say that an average crossover event affects 10 percent of the genes. Half of the events are deletions, which do not duplicate any genes, so that each event duplicates about 5 percent of the genes. We shall therefore take twenty crossover events as the equivalent of one generation. More generally, the number of events equivalent to one generation would be about $2/f$, where $f$ is the fraction of the genes duplicated or deleted in an average crossover event. Setting $G$, the number of genes, equal to its population genetics counterpart $2N$ (the number of chromosomes), we conclude that it requires about $2G$ generation-equivalents or $4G/f$ crossover events to completely turn over a family of $G$ genes.

It should be stressed that this is probably a gross overestimate of the turnover time. In ordinary population genetics it is assumed that the population breeds randomly, so that the probability that one particular chromosome will be propagated is independent of the probability that another will survive. This would certainly not be the case in the unequal crossover analogue. Recently duplicated descendants of one of the original genes would be clustered close together (though not, of course, necessarily contiguously) and thus tend to be duplicated or deleted together. The actual turnover may therefore be even orders of magnitude more rapid than the above estimate.

Nevertheless, our upper estimate indicates that a family the size of the rDNA system could turn over in a biologically reasonable time. If we take the above-mentioned figure of 10 percent for $f$, five hundred rDNA genes could be turned over by about twenty thousand crossing over events. If we allow one such crossover per ten meioses, these genes would turn over in about two hundred thousand years in an animal with a one-year generation time. The actual time may be much shorter, both because of our overestimation of the turnover time and because such unequal crossing over may not be restricted to meiosis but might also occur in the many germline mitoses by sister-strand crossover.

In putting forward this last proposal—that unequal crossover may occur predominantly by sister chromatid exchange during mitosis—I am actually making a virtue of necessity. Several lines of evidence argue against the obvious hypothesis that normal meiotic crossing over effects the proposed gene turnover. In the fruit fly, the *bb* (bobbed) "mutation" is thought to result from the deletion of some of the rDNA genes. Schalet (1969) examined partial revertants of *bb*, and all six which were unequivocally more like wild-type than the parental *bb* strains (and whose rDNA genes had therefore presumably increased in number) proved not to be recombinant for markers on opposite sides of the *bb* locus. On the other hand, when he examined about fifty recombinants which (because of the nature of the recombining chromosomes used) had to result from exchanges within the *bb* locus, almost all proved to have an altered *bb* genotype, whether more like wild-type or more extremely *bb* than the parental strains. Hence, while interchromosomal meiotic exchange can give rise to altered numbers of rDNA genes in this system, most such changes in number seem to arise in another manner. Moreover, one of the explanations of the rabbit $V_H$-region allotypes discussed in Chapter 7 required that gene turnover be predominantly intrachromosomal. Lastly, even admitting that the above estimate of the turnover time is much too high, the unequal crossover theory would be hard pressed to explain the complete turnover of twenty-five thousand 5sDNA genes while each gene accumulates few or no mutations if confined to meiotic exchanges.

If the turnover time is short compared to the time it takes for one of the repeated genes to acquire one mutation, the genes will be extremely homogeneous regardless of what selective forces are operating. Most mutations occurring in such genes will be rapidly deleted; but mutations will on rare occasions become predominant. Except for the short transition period of the turnover time, these will be the only two states of the system: essentially all genes unmutated (the usual case), or essentially all the genes harboring a single mutation (the rare case). Viewed on a broad evolutionary scale, it would seem that all the genes stay constant for prolonged periods of time but occasionally undergo a sudden simultaneous mutation.

If one of the repeated genes undergoes a mutation in a portion which is under selective pressure not to mutate, it will not survive for long. Either the vagaries of random unequal crossovers will eliminate it, or natural selection will eliminate it when its descendants come to represent a large portion of the genes in some ill-starred chromosomes.

On the other hand, mutations will accumulate in unselected portions of the repeated gene just as rapidly as if there were but a single gene. For in this case, the survival of any one gene through multiple crossover events is quite independent of whether or not it harbors mutations. These crossover events therefore can in no way influence the number of mutations which occur in the line of descent leading to any one contemporary gene: such a gene will accumulate mutations precisely as fast as if it descended through an equally ancient lineage of single genes.

The virtue of the crossover model is that it does not posit an unreasonably fastidious selective pressure on individual genes in a redundant family. The theory allows mutations to accumulate at random over a period of time approximately equal to one turnover time. It follows from this very attractive feature that all portions of the repeated unit in such a family should be equally homogeneous. If on the contrary the spacer portions of the repeated units accumulated several mutations during one turnover time (and became to that extent heterogeneous) while the transcribed parts accumulated none and remained homogeneous, it would imply that selection had eliminated mutations in the latter soon after they arose, before they could come to be represented in a large fraction of the repeated genes.

The uniform homogeneity of the repeated unit of the rDNA system thus fits the expectations of the crossover model. In this system the spacer portion of the repeated unit accumulates mutations rapidly in evolutionary time, yet is just as homogeneous as the highly conserved transcribed portion. The 5sDNA system, however, is at least superficially at variance with the expectations of the hypothesis, for in *Xenopus laevis* the spacer portion seems to be much more heterogeneous than the transcribed portion. The theory can be rescued if, as proposed above, the twenty-five thousand 5sDNA genes are distributed into several more or less independently evolving loci in this species. Whether or not this is so can probably be ascertained experimentally.

As explained in Chapter 7, gene turnover resulting from unequal crossover is a very plausible explanation for the evolution of the so-called species-specific differences, and even allotypic differences, in immunoglobulin V regions. Because these genes are heterogeneous, however, we must suppose that the turnover time is so long that one gene suffers multiple mutations during it. I have not proved that the observed distribution of V genes into subgroups is expected under these circumstances. But this outcome is plausible since it seems likely that rather late in the course of one turnover time an intermediate state is reached in which descendants of only a few of the original genes remain, and in which all the surviving descendants of any one of the original genes have only recently diverged from one another.

The immunoglobulin V regions, however, pose a difficulty for the crossover theory similar to that posed by the 5sDNA system, for all portions of V genes are not equally heterogeneous. As mentioned in Chapter 5, some of the positions in these sequences are nearly invariant while others are hypervariable. If, as proposed in the crossover theory, mutations accumulate at random in these genes, the chance that they will assume this pattern is negligible, and we can conclude tentatively that selection is eliminating deleterious mutations in the invariant positions (and perhaps amplifying mutations in the hypervariable positions) well before they come to be represented in a large fraction of the genes. This inference by no means implies that V genes do not turn over by unequal crossing over; it simply means that this turnover does not obviate the necessity (embarrassing to proponents of the germline theory of diversity) to suppose that V genes are under rather close surveillance by natural selection.

In short, V genes do not seem to be as redundant as the immunoglobulins themselves. The simplest conclusion from these considerations is that there are only relatively few, highly selected germline V genes, and that the apparent redundancy of immunoglobulins arises from superimposed somatic mutations. As mentioned in Chapter 5, however, this is by no means a certain conclusion and it seems wise to await more direct evidence before making a final decision between the germline and somatic theories.

The unequal crossover theory is uneconomical in that it does not explain why multiple genes occur as powers of two. It should be kept in mind, however, that there is no necessary connection between this pervasive power series and the evolution of such families as if they were single genes. There is ample room in the data for small fluctuations in gene number due to unequal crossing over superimposed on the more obvious factor-of-two changes which cannot have that explanation.

The essential feature of the unequal crossover theory's explanation for the conservation of redundant genes is that this crossover will lead to gene turnover. In Smithies' branched network (Chapter 9), branch formation and branch elimination have the same effect. Moreover, these processes are necessarily intrachromosomal, and the evidence against meiotic exchange as the turnover mechanism is thus an expectation rather than a complication of the theory.

### The Master-Slave Hypothesis

According to Callan's (1967) master-slave hypothesis, each gene in a redundant multigene family is matched against a single "master" gene at every meiosis and corrected if it differs. The entire family evolves exactly in parallel with the master copy, all mutations except those in the master copy being suppressed at the next meiosis.

The cytological evidence Callan cites in favor of his theory has no very direct connection with its central features, and I find little else to recommend it. Even more strongly than the crossover theory, it predicts that all portions of the repeated unit are equally homogeneous, and it thus has no advantage over the crossover theory in explaining the puzzling heterogeneity of the 5sDNA spacer. It invokes a special correction mechanism to explain facts which the well-precedented process of unequal exchange explains at least as well. And it does not account for the evolution of V genes at all, since it does not accommodate diversity within the slave genes.

Akin to the master-slave hypothesis is the democratic gene conversion model of Gally and Edelman, which was described in Chapter 5. Their theory differs from that of Callan in that there is no master gene, the correction is random, and each of the gene-conversion events affects only a portion of the repeated sequence. It has the virtue of accommodating diversity within the family and thus applying to antibodies. In addition, it can accommodate the observed nonuniform heterogeneity along the repeated sequence of both 5sDNA and V-gene systems. As mentioned in Chapter 5, however, the abnormally high incidence of dislocations it predicts is not in fact observed among the immunoglobulins.

# References


Abdou, N.I., and M. Richter (1969). "Cells involved in the immune response. X. The transfer of antibody-forming capacity to irradiated rabbits by antigen-reactive cells isolated from normal allogeneic rabbit bone marrow after passage through antigen-sensitized glass bead columns," *J. Exp. Med.* 130:141.

Abel, C.A., and H.M. Grey (1967). "Carboxy-terminal amino acids of γA and γM heavy chains," *Science* 156:1609.

—— —— (1968). "Studies on the structure of mouse γA myeloma proteins," *Biochemistry* 7:2672.

Acton, R.T., P.F. Weinheimer, S.J. Hall, W. Niedermeier, E. Shelton, and J.C. Bennett (1971). "Tetrameric immune macroglobulins in three orders of bony fishes," *Proc. Nat. Acad. Sci. U.S.A.* 68:107.

Ada, G.L. (1970). "Antigen binding cells in tolerance and immunity," *Transpl. Rev.* 5:105.

Adler, F.L., M. Fishman, and S. Dray (1966). "Antibody formation initiated in vitro. III. Antibody formation and allotypic specificity directed by ribonucleic acid from peritoneal exudate cells," *J. Immunol.* 97:554.

Aloni, Y., L.E. Hatlen, and G. Attardi (1971). "Studies of fractionated HeLa cell metaphase chromosomes. II. Chromosomal distribution of sites for transfer RNA and 5*s* RNA," *J. Mol. Biol.* 56:555.

Appella, E. (1971). "Amino acid sequences of two mouse immunoglobulin lambda chains," *Proc. Nat. Acad. Sci. U.S.A.* 68:590.

—— A. Chersi, R.G. Mage, and S. Dubiski (1971). "Structural basis of the *A* 14 and *A* 15 allotypic specificities in rabbit immunoglobulin G," *Proc. Nat. Acad. Sci. U.S.A.* 68:1341.

—— —— J. Rejnek, and R.A. Reisfeld (1970). "Studies on the amino acid sequence of rabbit light chains," in *Developmental Aspects of Antibody Formation and Structure* (Prague, Academia), p. 327.

—— and D. Ein (1967). "Two types of lambda polypeptide chains in human immunoglobulin based on an amino acid substitution at position 190," *Proc. Nat. Acad. Sci. U.S.A.* 57:1449.

—— R.G. Mage, S. Dubiski, and R.A. Reisfeld (1968). "Chemical and immunochemical evidence for different classes of rabbit light polypeptide chains," *Proc. Nat. Acad. Sci. U.S.A.* 60:975.

—— J. Rejnek, and R.A. Reisfeld (1969). "Variations of the carboxy-terminal amino acid sequence of rabbit light chain with *b*4, *b*5, and *b*6 allotypic specificities," *J. Mol. Biol.* 41:473.

Attardi, G., M. Cohn, K. Horibata, and E.S. Lennox (1959). "On the analysis of antibody synthesis at the cellular level," *Bact. Rev.* 23:213.

—— —— —— —— (1964). "Antibody formation by rabbit lymph node cells. I. Single cell responses to several antigens," *J. Immunol.* 92:335.

Avey, H.P., R.J. Poljak, G. Rossi, and A. Nisonoff (1968). "Crystallographic data for the Fab fragment of a human myeloma immunoglobulin," *Nature* 220:1248.

Bach, F.H., H. Bock, K. Graupner, E. Day, and H. Klostermann (1969). "Cell kinetic studies in mixed leukocyte cultures: an in vitro model of homograft reactivity," *Proc. Nat. Acad. Sci. U.S.A.* 62:377.

Baczko, K., D.G. Braun, M. Hess, and N. Hilschmann (1970). "Die Primaerstruktur einer monoklonalen immunoglobulin L-Kette der Subgruppe IV vom λ-Typ (Bence-Jones-Protein Bau): Untergruppen innerhalb der Subgruppen," *Z. Phys. Chem.* 351:763.

Baglioni, C., L. Alescio-Zonta, D. Cioli, and A. Carbonara (1966). "Allelic antigenic factor InV(a) of the light chains of human immunoglobulins: chemical basis," *Science* 152:1517.

Bauer, D.C., M.J. Mathies, and A.B. Stavitsky (1963). "Sequences of synthesis of γ-1 macroglobulin and γ-2 globulin antibodies during primary and secondary responses to proteins, *Salmonella* antigens, and phage," *J. Exp. Med.* 117:889.

Beale, D., and M. Squires (1970). "C-terminal amino acid sequence of bovine immunoglobulin light chain," *Nature* 226:1056.

Bell, C., and S. Dray (1969). "Conversion of non-immune spleen cells by ribonucleic acid of lymphoid cells from an immunized rabbit to produce γM antibody of foreign light chain allotype," *J. Immunol.* 103:1196.

—— —— (1971). "Expression of allelic immunoglobulin in rabbits injected with RNA extract," *Science* 171:199.

Bernier, G.M., R.E. Ballieux, K.T. Tominaga, and F.W. Putnam (1967). "Heavy chain subclasses of human γG-globulin. Serum distribution and cellular localization," *J. Exp. Med.* 125:303.

—— and J.J. Cebra (1965). "Frequency distribution of α, γ, κ, and λ polypeptide chains in human lymphoid tissues," *J. Immunol.* 95:246.

Bosma, M., and E. Weiler (1970). "The clonal nature of antibody formation. I. Clones of antibody-forming cells of poly-D-alanine specificity," *J. Immunol.* 104:203.

Bourgois, A., and M. Fougereau (1970). "Isolation and characterization of the cyanogen bromide fragments of a mouse γG immunoglobulin," *Eur. J. Biochem.* 12:558.

Bozzi, G., R.A. Binaghi, C. Stiffel, and D. Monton (1969). "Production of different classes of immunoglobulin by individual cells in the guinea-pig," *Immunology* 16:349.

Bradshaw, C.M., L.W. Clem, and M.M. Sigel (1969). "Immunologic and immunochemical studies on the gar, *Lepisosteus platyrhincus*," *J. Immunol.* 103:496.

Breinl, F., and F. Haurowitz (1930). "Chemische Untersuchungen des Praezipitates aus Haemoglobin und Antihaemoglobin-Serum und Bemerkungen ueber die Natur der Antikoerper," *Z. Phys. Chem.* 192:45.

Brenner, S., and C. Milstein (1966). "Origin of antibody variation," *Nature* 211:242.

Britten, R.J., and D.E. Kohne (1968). "Repeated sequences in DNA," *Science* 161:529.

Brown, D.D. (1971). "The evolutionary solution to the antibody problem," presented at the September 4 symposium of the Society of General Physiologists.

—— and A.W. Blackler (1972). "Gene amplification proceeds by a chromosome copy mechanism," *J. Mol. Biol.* 63:75.

—— and I.B. Dawid (1969). "Developmental Genetics," *Ann. Rev. Genetics* 3:127.

—— and C.S. Weber (1968). "Gene linkage by RNA-DNA hybridization. I. Unique DNA sequences homologous to 4*s*RNA, 5*s*RNA, and ribosomal RNA. II. Arrangement of the redundant gene sequences for 28*s* and 18*s* ribosomal RNA," *J. Mol. Biol.* 34:661 and 681.

—— P.C. Wensink, and E. Jordan (1971). "Purification and some characteristics of 5*S* DNA from *Xenopus laevis*," *Proc. Nat. Acad. Sci. U.S.A.* 68:3175.

—— —— —— (1972). "A comparison of the ribosomal DNA's of *Xenopus laevis* and *Xenopus mulleri*: the evolution of tandem genes," *J. Mol. Biol.* 63:57.

Buongiorno-Nardelli, M., F. Amaldi, and P.A. Lava-Sanchez (1972). "Amplification as a rectification mechanism for the redundant rRNA genes," *Nature New Biol.* 238:134.

Burnet, F.M. (1959). *The Clonal Selection Theory of Acquired Immunity* (Nashville, Tenn., Vanderbilt University Press).

—— (1967). "The impact on ideas of immunology," *Cold Spr. Hbr. Symp.* 32:1.

Callan, H.G. (1967). "The organization of genetic units in chromosomes," *J. Cell Science* 2:1.

Capra, J.D., and H.G. Kunkel (1970). "Amino acid sequence similarities in two human anti-gamma globulin antibodies," *Proc. Nat. Acad. Sci. U.S.A.* 67:87.

Cebra, J.J., B. Birshtein, Q. Hussain, and K.J. Turner (1970). "The structure of IgG(2) and the chemical basis of antibody specificity: the approach based on normal immunoglobulin and specific antibodies," in *Developmental Aspects of Antibody Formation and Structure* (Prague, Academia), p. 363.

—— J.E. Coldberg, and S. Dray (1966). "Rabbit lymphoid cells differentiated with respect to $\alpha$-, $\gamma$-, and $\mu$-heavy polypeptide chains and to allotypic markers *Aa*1 and *Aa*2," *J. Exp. Med.* 123:547.

Chersi, A., R.G. Mage, J. Rejnek, and R.A. Reisfeld (1970). "Isolation,

chemical and immunological characterization of κ- and λ-type light chains from IgG of normal rabbits with *b*9 allotype," *J. Immunol.* 104:1205.
Chesebro, B., B. Bloth, and S.E. Svehag (1968). "The ultrastructure of normal and pathological IgM immunoglobulins," *J. Exp. Med.* 127:399.
Chiller, J.M., G.S. Habicht, and W.O. Weigle (1970). "Cellular sites of immunological unresponsiveness," *Proc. Nat. Acad. Sci. U.S.A.* 65:551.
Claman, H.N., E.A. Chaperon, and R.F. Triplett (1966). "Thymus-marrow cell combinations. Synergism in antibody production," *Proc. Soc. Exp. Med.* 122:1167.
Cohen, E.P. (1967). "The appearance of new species of RNA in the mouse spleen after immunization as detected by molecular hybridization," *Proc. Nat. Acad. Sci. U.S.A.* 57:673.
—— and K. Raska (1968). "Unique species of RNA in peritoneal cells exposed to different antigens," in *Nucleic Acids in Immunology* (New York, Springer-Verlag), p. 573.
Cohn, M. (1968). "The molecular biology of expectation," in *Nucleic Acids in Immunology* (New York, Springer-Verlag), p. 671.
—— (1970). "Selection under a somatic model," *Cell. Immunol.* 1:461.
Conway, T.P., S. Dray, and E.A. Lichter (1969a). "Identification and genetic control of three rabbit γA immunoglobulin allotypes," *J. Immunol.* 102:544.
—— —— —— (1969b). "Identification and genetic control of the *f*4 and *f*5 rabbit γA immunoglobulin allotypes," *J. Immunol.* 103:662.
Cooper, A.G., and A.G. Steinberg (1970). "INV allotypes of cold agglutinin kappa chains," *J. Immunol.* 104:1108.
Cosenza, H., and A.A. Nordin (1970). "Immunoglobulin classes of antibody-forming cells in mice," *J. Immunol.* 104:976.
Cudkowicz, G., G.M. Shearer, and R.L. Priore (1969). "Cellular differentiation of the immune system of mice. V. Class differentiation in marrow precursors of plaque-forming cells," *J. Exp. Med.* 130:481.
Cunningham, B.A., M. Pflumm, U. Rutishauer, and G.M. Edelman (1969). "Subgroups of amino acid sequences in the variable regions of immunoglobulin heavy chains," *Proc. Nat. Acad. Sci. U.S.A.* 64:997.
Daugharty, H., J.E. Hopper, A.B. MacDonald, and A. Nisonoff (1969). "Quantitative investigation of idiotypic antibodies. I. Analysis of precipitating antibody populations," *J. Exp. Med.* 130:1047.
David, G.S., and C.W. Todd (1969). "Suppression of heavy and light chain allotypic expression in homozygous rabbits through embryo transfer," *Proc. Nat. Acad. Sci. U.S.A.* 62:860.
Davie, J.M., W.E. Paul, R.G. Mage, and M.B. Goldman (1971). "Membrane-associated immunoglobulin of rabbit peripheral blood lymphocytes: allelic exclusion at the *b* locus," *Proc. Nat. Acad. Sci. U.S.A.* 68:430.
Davis, B.D., R. Dulbecco, H.N. Eisen, H.S. Ginsberg, and W.B. Wood (1967). *Microbiology* (New York, Hoeber).
Dayhoff, M.O., ed. (1969). *Atlas of Protein Sequence and Structure*

(Silver Spring, Md., National Biomedical Research Foundation).

dePreval, C., J.R.L. Pink, and C. Milstein (1970). "Variability of interchain binding of immunoglobulins. Interchain bridges of mouse $IgG_{2a}$ and $IgG_{2b}$," *Nature* 228:930.

Doolittle, R.F., and K.H. Astrin (1967). "Light chains of rabbit immunoglobulin: assignment to the kappa class," *Science* 156:1755.

Dorrington, K.J., and C. Tanford (1970). "Molecular size and conformation of immunoglobulins," *Adv. Immunol.* 12:333.

—— M.H. Zarlengo, and C. Tanford (1967). "Conformational change and complementarity in the combination of H and L chains of immunoglobulin-G," *Proc. Nat. Acad. Sci. U.S.A.* 58:996.

Dray, S. (1962). "Effect of maternal isoantibodies on the quantitative expression of two allelic genes controlling γ-globulin allotypic specificities," *Nature* 195:677.

—— G.O. Young, and A. Nisonoff (1963). "Distribution of allotypic specificities among rabbit γ-globulin molecules genetically defined at two loci," *Nature* 199:52.

Dreyer, W.J. (1970). "A proposed new and general chromosomal control mechanism for commitment of specific cell lines during development," in *Developmental Aspects of Antibody Formation and Structure* (Prague, Academia), p. 919.

—— and J.C. Bennett (1965). "The molecular basis of antibody formation: a paradox," *Proc. Nat. Acad. Sci. U.S.A.* 54:864.

—— and W.R. Gray (1968). "On the role of nucleic acids as genes conferring precise chemospecificity to differentiated cell lines," in *Nucleic Acids in Immunology* (New York, Springer-Verlag), p. 614.

—— —— and L. Hood (1967). "The genetic, molecular, and cellular basis of antibody formation: some facts and a unifying hypothesis," *Cold Spr. Hbr. Symp.* 32:353.

Dubiski, S. (1967a). "Synthesis of allotypically defined immunoglobulins in rabbits," *Cold Spr. Hbr. Symp.* 32:311.

—— (1967b). "Suppression of synthesis of allotypically defined immunoglobulins and compensation by another subclass of immunoglobulin," *Nature* 214:1365.

—— (1969). "Immunochemistry and genetics of a new allotypic specificity *Ae*14 of rabbit γG immunoglobulins: recombination in somatic cells," *J. Immunol.* 103:120.

—— and P.J. Muller (1967). "A 'new' allotypic specificity (*A*9) of rabbit immunoglobulin," *Nature* 214:696.

Dutton, R.W., and R.I. Mishell (1967). "Cell population and cell proliferation in the *in vitro* response of normal mouse spleen to heterologous erythrocytes. Analysis by the hot pulse technique," *J. Exp. Med.* 126:443.

Edelman, G.M., B.A. Cunningham, W.E. Gall, P.D. Gottlieb, U. Rutischauer, and M.J. Waxdal (1969). "The covalent structure of an entire γG immunoglobulin molecule," *Proc. Nat. Acad. Sci. U.S.A.* 63:78.

—— and J.A. Gally (1970). "Arrangement and evolution of eukaryotic genes," in *The Neurosciences: Second Study Program*, ed. F.O. Schmitt (New York, Rockefeller University Press), p. 962.
—— and P.D. Gottlieb (1970). "A genetic marker in the variable region of light chains of mouse immunoglobulins," *Proc. Nat. Acad. Sci. U.S.A.* 67:1192.
Edman, P., and A.G. Cooper (1968). "Amino acid sequence at the N-terminal end of a cold agglutinin kappa chain," *Fed. Eur. Biochem. Soc. Letters* 2:33.
Edmundson, A.B., R.M. Rennebohm, T. Johnson, K.R. Ely, F.A. Sheber, M. Schiffer, M.K. Wood, and O. Smithies (1971). Unpublished observation.
—— F.A. Sheber, K.R. Ely, N.B. Simonds, N.K. Hutson, and J.L. Rossiter (1968). "Characterization of human L type Bence-Jones proteins containing carbohydrate," *Arch. Biochem. Biophys.* 127:725.
Ehrlich, P. (1900). "On immunity with special reference to cell life," *Proc. Royal Soc.*, ser. B, 66:424.
Eichmann, K., D.G. Braun, T. Feizi, and R.M. Krause (1970). "The emergence of antibodies with either identical or unrelated individual antigenic specificity during repeated immunization with streptococcal vaccines," *J. Exp. Med.* 131:1169.
Ein, D. (1968). "Nonallelic behavior of the Oz groups in human lambda immunoglobulin chains," *Proc. Nat. Acad. Sci. U.S.A.* 60:982.
—— and J. Fahey (1967). "Two types of lambda polypeptide chains in human immunoglobulins," *Science* 156:947.
Feingold, I., J.L. Fahey, and T.F. Dutcher (1968). "Immunofluorescent studies of immunoglobulin in human lymphoid cells in continuous culture," *J. Immunol.* 101:366.
Feinstein, A. (1963). "Character and allotypy of an immune globulin in rabbit colostrum," *Nature* 199:1197.
—— P.G.H. Gell, and A.S. Kelus (1963). "Immunochemical analysis of rabbit gamma-globulin allotypes," *Nature* 200:653.
—— and E.A. Munn (1969). "Conformation of the free and antigen-bound IgM antibody molecules," *Nature* 224:1307.
Fisher, C.E., W.H. Palm, and E.M. Press (1969). "The N-terminal sequence of a human $\gamma_1$ chain of allotype *Gm*(a-f+)," *Fed. Eur. Biochem. Soc. Letters* 5:20.
Fitch, W.M. (1966). "An improved method of testing for evolutionary homology," *J. Mol. Biol.* 16:9.
—— and E. Margoliash (1967). "Construction of phylogenetic trees: a method based on mutation distances as estimated from cytochrome c is of general applicability," *Science* 155:279.
—— —— (1968). "The construction of phylogenetic trees. II. How well do they reflect past history?" *Brookhaven Symposium in Biology* 21:217.
Fleischman, J.B. (1967). "Synthesis of the γG heavy chain in rabbit lymph node cells," *Biochemistry* 6:1311.

Francis, T.C., and W.E. Paul (1970). "Inhibition by hapten of cellular immune responses to a hapten-protein conjugate," *Nature* 226:173.

Franek, F., B. Keil, and F. Sorm (1970). "Amino acid sequence of a 22-residue section from the variable part of pig immunoglobulin λ-chains," *Eur. J. Biochem.* 11:170.

—— and J. Novotny (1969). "Isolation of a disulfide containing fragment of the variable part of pig immunoglobulin λ-chains," *Eur. J. Biochem.* 11:165.

Frangione, B. (1970). "Correlation of the C-terminal sequence of rabbit light chains with allotypes," *Fed. Eur. Biochem. Soc. Letters* 3:341.

—— E.C. Franklin, H.H. Fudenberg, and M.E. Koshland (1966). "Structural studies of human γG-myeloma proteins of different antigenic subgroup and genetic specificities," *J. Exp. Med.* 124:715.

—— —— and A.S. Kelus (1968). "L-chain peptide maps of rabbit IgG of different allotypes," *Immunology* 15:599.

—— and C. Milstein (1968). "Variations in the S-S bridges of immunoglobulins G: interchain disulfide bridges of $\gamma G_3$ myeloma proteins," *J. Mol. Biol.* 33:893.

—— —— (1969). "Partial deletion in the heavy chain disease protein Zuc," *Nature* 224:597.

—— —— and J.R.L. Pink (1969). "Structural studies of immunoglobulin G," *Nature* 221:145.

Freedman, M.H., and M. Sela (1966). "Recovery of specific activity upon reoxidation of completely reduced polyalanyl rabbit antibody," *J. Biol. Chem.* 241:5225.

Fruchter, R.G., S.A. Jackson, L.E. Mole, and R.R. Porter (1970). "Sequence studies of the Fd section of the heavy chain of rabbit immunoglobulin G," *Biochem. J.* 116:249.

Fudenberg, H.H., G. Drews, and A. Nisonoff (1964). "Serologic demonstration of dual specificity of rabbit bivalent hybrid antibody," *J. Exp. Med.* 119:151.

Gally, J.A., and G.M. Edelman (1970). "Somatic translocation of antibody genes," *Nature* 227:341.

Gershon, H., S. Baumberger, M. Sela, and M. Feldman (1968). "Studies on the competence of single cells to produce antibodies of two specificities," *J. Exp. Med.* 128:223.

Gibson, D., M. Levanon, and O. Smithies (1971). "Heterogeneity of normal human immunoglobulin light chains. Nonallelic variation in the constant region of lambda chains," *Biochemistry* 10:3114.

Gilman, A.M., A. Nisonoff, and S. Dray (1964). "Symmetrical distribution of genetic markers in individual rabbit γ-globulin markers," *Immunochemistry* 1:109.

Gilman-Sachs, A., R.G. Mage, G.O. Young, C. Alexander, and S. Dray (1969). "Identification and genetic control of two rabbit immunoglobulin allotypes at a second light chain locus, the *c*-locus," *J. Immunol.* 103:1159.

Givol, D., and F. DeLorenzo (1968). "The position of various cleavages of rabbit immunoglobulin G," *J. Biol. Chem.* 243:1886.

Goetzl, E.J., and H. Metzger (1970). "Affinity labeling of a mouse myeloma protein which binds nitrophenyl ligands. Sequence and position of a labeled tryptic peptide," *Biochemistry* 9:3862.

Goldstein, D.J., R.L. Humphrey, and R.J. Poljak (1968). "Human Fc fragment: crystallographic evidence for two equivalent subunits," *J. Mol. Biol.* 35:247.

Gray, W.R., W.J. Dreyer, and L. Hood (1967). "Mechanism of antibody synthesis: size differences between mouse kappa chains," *Science* 155:465.

Greaves, M.F. (1970). "Biological effects of anti-immunoglobulins: evidence for immunoglobulin receptors on 'T' and 'B' lymphocytes," *Transpl. Rev.* 5:45.

Green, I., P. Vassalli, and B. Benacerraf (1967a). "Cellular localization of anti-DNP-PLL and anticonveyor albumin antibodies in genetic nonresponder guinea pigs immunized with DNP-PLL albumin complexes," *J. Exp. Med.* 125:527.

—— —— V. Nussenzweig, and B. Benacerraf (1967b). "Specificity of the antibodies produced by single cells following immunization with antigens bearing two types of antigenic determinants," *J. Exp. Med.* 125:511.

Green, N.M. (1969). "Electron microscopy of the immunoglobulins," *Adv. Immunol.* 11:1.

Greenberg, L.J., and J.W. Uhr (1967a). "DNA-RNA hybridization studies of myeloma tumors in mice," *Proc. Nat. Acad. Sci. U.S.A.* 58:1878.

—— —— (1967b). "DNA-RNA hybridization studies of immunoglobulin synthesizing tumors in mice," *Cold Spr. Hbr. Symp.* 32:243.

Grey, H.M. (1969a). "Phylogeny of immunoglobulins," *Adv. Immunol.* 10:51.

—— (1969b). "Presence of L-L interchain disulfide bonds in reconstituted $\gamma$G molecules," *J. Immunol.* 102:848.

—— C.A. Abel, W.J. Yount, and H.G. Kunkel (1968). "A subclass of human $\gamma$A-globulins ($\gamma A_2$) which lacks the disulfide bonds linking heavy and light chains," *J. Exp. Med.* 128:1223.

—— and M. Mannik (1965). "Specificity of recombination of H and L chains from human $\gamma$G-myeloma proteins," *J. Exp. Med.* 122:619.

—— A. Shei, and M. Shalitin (1970). "The subunit structure of mouse IgA," *J. Immunol.* 105:75.

Grubb, R. (1970). *The Genetic Markers of Human Immunoglobulins* (New York, Springer-Verlag).

Guillien, P., S. Avrameas, and P. Burtin (1970). "Specificity of antibodies in single cells after immunization with antigens bearing several antigenic determinants. A study with a new paired staining technique," *Immunology* 18:483.

Haber, E. (1964). "Recovery of antigenic specificity after denaturation

and complete reduction of disulfides in a papain fragment of antibody," *Proc. Nat. Acad. Sci. U.S.A.* 52:1099.
— (1968). "Immunochemistry," *Ann. Rev. Biochem.* 37:497.
— F.F. Richards, J. Spragg, K.F. Austen, M. Vallatton, and L.B. Page (1967). "Modifications in the heterogeneity of the antibody response," *Cold Spr. Hbr. Symp.* 32:299.
Halpern, M.S., and M.E. Koshland (1970). "Novel subunit in secretory IgA," *Nature* 228:1276.
Hamers, R., and C. Hamers-Casterman (1965). "Molecular localization of A chain allotypic specificities in rabbit IgG (7s γ-globulin)," *J. Mol. Biol.* 14:288.
— — (1967). "Evidence for the presence of the Fc allotype marker *As*8 and the Fd allotype marker *As*1 in the same molecules of rabbit IgG," *Cold Spr. Hbr. Symp.* 32:129.
— — and A.S. Kelus (1965). "Un gène nouveau intervenant dans la synthèse de la γ-globuline du lapin," *Arch. Int. Physiol. Biochem.* 73:147.
— — and S. Lagnaux (1966). "A new allotype in the rabbit linked with *As*1 which may characterize a new class of IgG," *Immunology* 10:399.
Haurowitz, F. (1967). "The evolution of selective and instructive theories of antibody formation," *Cold Spr. Hbr. Symp.* 32:559.
Henney, C.S., H.D. Welscher, W.D. Terry, and D.S. Rowe (1969). "Studies on human IgD. II. The lack of skin sensitizing and complement fixing activities of immunoglobulin D," *Immunochemistry* 6:445.
Herzenberg, L.A., H.O. McDevitt, and L.A. Herzenberg (1968). "Genetics of antibodies," *Ann. Rev. Genetics* 2:209.
— J.D. Minna, and L.A. Herzenberg (1967). "The chromosome region for immunoglobulin heavy chains in the mouse: allelic electrophoretic mobility differences and allotype suppression," *Cold Spr. Hbr. Symp.* 32:181.
Hess, M., and N. Hilschmann (1970). "Genetische Polymorphismus im Konstanten Teil von humanen Immunoglobulin-L-Ketten vom λ-Typ," *Z. Phys. Chem.* 351:67.
Hill, R.L., R. Delaney, R.E. Fellows, and H.E. Lebovitz (1966). "The evolutionary origins of the immunoglobulins," *Proc. Nat. Acad. Sci. U.S.A.* 56:1762.
Hilschmann, N. (1969). "Die molekularen Grundlagen der Antikoerperbildung," *Naturwissenschaften* 56:195.
— and L.C. Craig (1965). "Amino acid sequence studies with Bence-Jones proteins," *Proc. Nat. Acad. Sci. U.S.A.* 53:1403.
Hood, L., K. Eichmann, H. Lackland, R.M. Krause, and J.J. Ohms (1970). "Rabbit antibody light chains and gene evolution," *Nature* 228:1040.
— and D. Ein (1968a). "Genetic implications of common region sequence comparisons of lambda immunoglobulin chains differing at position 190," *Science* 162:679.

—— —— (1968b). "Immunoglobulin lambda chain structure: two genes, one polypeptide," *Nature* 220:764.

—— J.A. Grant, and H.C. Sox (1970). "On the structure of normal light chains from mammals and birds: evolutionary and genetic implications," in *Developmental Aspects of Antibody Formation and Structure* (Prague, Academia), p. 283.

—— W.R. Gray, B.G. Sanders, and W.J. Dreyer (1967). "Light chain evolution," *Cold Spr. Hbr. Symp.* 32:133.

—— D. McKean, V. Farnsworth, and M. Potter (1972). In preparation.

—— M. Potter, and D. McKean (1971). "Immunoglobulin structure: amino terminal sequences of kappa chains from genetically similar mice (BALB/c)," *Science* 170:1207.

—— and D.W. Talmage (1970). "Mechanism of antibody diversity: germ line basis for variability," *Science* 168:325.

Hopper, J.E., A.B. MacDonald, and A. Nisonoff (1970). "Quantitative investigation of idiotypic antibodies. II. Nonprecipitating antibodies," *J. Exp. Med.* 131:41.

Humphrey, R.L. (1967). "Crystallographic study of human myeloma Fc-fragment," *J. Mol. Biol.* 29:525.

—— H.P. Avey, L.N. Becka, R.J. Poljak, G. Rossi, T.K. Choi, and A. Nisonoff (1969). "X-ray crystallographic study of the Fab fragments from two human myeloma proteins," *J. Mol. Biol.* 43:223.

Inman, J.K., and R.A. Reisfeld (1968). "Differences in amino acid composition of papain Fd fragments from rabbit γG-immunoglobulins carrying different H chain allotypic specificities," *Immunochemistry* 5:415.

Ishizaka, K., and T. Ishizaka (1967). "Identification of γE-antibodies as a carrier of reaginic activity," *J. Immunol.* 99:1187.

Jacobson, E.B., J. L'age-Stehr, and L.A. Herzenberg (1970). "Immunological memory in mice. II. Cell interaction in the secondary immune response studied by means of immunoglobulin allotype markers," *J. Exp. Med.* 131:1109.

Jaton, J.C., N.R. Klinman, D. Givol, and M. Sela (1968). "Recovery of antibody activity upon reoxidation of completely reduced polyalanyl heavy chain and its Fd fragment derived from anti-2,4-dinitrophenyl antibody," *Biochemistry* 7:4185.

—— M.D. Waterfield, M.N. Margolies, and E. Haber (1970). "Isolation and characterization of structurally homogeneous antibodies from antipneumococcal sera," *Proc. Nat. Acad. Sci. U.S.A.* 66:959.

Jerne, N.K. (1970). "The somatic generation of immune recognition," *Eur. J. Immunol.* 1:1.

Jerry, L.M., H.G. Kunkel, and H.M. Grey (1970). "Absence of disulfide bonds linking the heavy and light chains: a property of a genetic variant of $\gamma A_2$ globulins," *Proc. Nat. Acad. Sci. U.S.A.* 65:557.

Kaplan, A.P., and H. Metzger (1969). "Partial sequences of six macro-

globulin light chains. Absence of sequence correlates to functional activity," *Biochem.* 8:3944.
Katz, D.H., W.E. Paul, E.A. Goidl, and B. Benacerraf (1970a). "Radio-resistance of the cooperative function of carrier-specific lymphocytes in anti-hapten antibody responses," *Science* 170:462.
—— —— —— —— (1970b). "Carrier function in anti-hapten antibody responses. I. Enhancement of primary and secondary antibody responses by carrier preimmunization. II. Specific properties of carrier cells capable of enhancing anti-hapten antibody responses," *J. Exp. Med.* 132:261 and 283.
—— —— —— —— (1971). "Carrier function in anti-hapten antibody responses. III. Stimulation of antibody synthesis and facilitation of hapten-specific secondary responses by graft-versus-host reactions," *J. Exp. Med.* 133:169.
Kelus, A.S., and P.G.H. Gell (1965). "An allotypic determinant specific to rabbit macroglobulin," *Nature* 206:313.
Kincade, P.W., A.R. Lawton, D.E. Bockman, and M.D. Cooper (1970). "Suppression of immunoglobulin G synthesis as a result of antibody-mediated suppression of immunoglobulin M synthesis in chickens," *Proc. Nat. Acad. Sci. U.S.A.* 67:1918.
Kindt, T.J., M.W. Steward, and C.W. Todd (1969). "Allotypic markers on rabbit IgA," *Biochem. Biophys. Res. Comm.* 31:9.
—— and C.W. Todd (1969). "Heavy and light chain allotypic markers on rabbit homocytotropic antibody," *J. Exp. Med.* 130:859.
Knopf, P.M., R.M.E. Parkhouse, and E.S. Lennox (1967). "Biosynthetic units of an immunoglobulin heavy chain," *Proc. Nat. Acad. Sci. U.S.A.* 58:2288.
Kohler, H., A. Shimizu, C. Paul, V. Moore, and F.W. Putnam (1970a). "Three variable-gene pools common to IgM, IgG, and IgA immunoglobulins," *Nature* 227:1318.
—— —— —— and F.W. Putnam (1970b). "Macroblobulin structure: variable sequence of light and heavy chains," *Science* 169:56.
Koshland, M.E. (1967). "Location of specificity and allotypic amino acid residues in antibody Fd fragments," *Cold Spr. Hbr. Symp.* 32:119.
—— J.J. Davis, and N.J. Fugita (1969). "Evidence for multiple gene control of a single polypeptide chain: the heavy chain of rabbit immunoglobulin," *Proc. Nat. Acad. Sci. U.S.A.* 63:1274.
—— and F.M. Englberger (1963). "Differences in the amino acid composition of two purified antibodies from the same rabbit," *Proc. Nat. Acad. Sci. U.S.A.* 50:61.
—— —— and R. Shapanka (1966). "Location of amino acid differences in the subunits of three rabbit antibodies," *Biochemistry* 5:641.
—— R.A. Reisfeld, and S. Dray (1968). "Differences in amino acid composition related to allotypic and antibody specificity of rabbit IgG heavy chains," *Immunochemistry* 5:471.

Kubo, R.T., I.Y. Rosenblum, and A.A. Benedict (1970). "The unblocked N-terminal sequence of chicken λ-like light chains," *J. Immunol.* 105:534.
Kunkel, H.G., F.G. Joslin, G.M. Penn, and J.B. Natvig (1970). "Genetic variants of $\gamma G_4$ globulin. A unique relationship to other classes of γG globulin," *J. Exp. Med.* 132:508.
—— J.B. Natvig, and F.G. Joslin (1969). "A 'Lepore' type of hybrid γ globulin," *Proc. Nat. Acad. Sci. U.S.A.* 62:144.
L'age-Stehr, J., and L.A. Herzenberg (1970). "Immunological memory in mice. I. Physical separation and partial characterization of memory cells for different immunoglobulin classes from each other and from antibody-producing cells," *J. Exp. Med.* 131:1093.
Landsteiner, K. (1945). *The Specificity of Serological Reactions* (Cambridge, Mass., Harvard University Press).
Langer, B., M. Steinmetz-Kayne, and N. Hilschmann (1968). "Die vollstaendig aminosaeuresequenz des Bence-Jones-Proteins (λ Typ). Subgruppen im variablen Teil bei Immunoglobulin-L-Ketten vom λ-Typ," *Z. Phys. Chem.* 349:945.
Lennox, E.S., P.M. Knopf, A.J. Munro, and R.M.E. Parkhouse (1967). "A search for biosynthetic subunits of light and heavy chains of immunoglobulins," *Cold Spr. Hbr. Symp.* 32:249.
Lesley, J., and R.W. Dutton (1970). "Antigen receptor molecules; inhibition by antiserum against kappa light chains," *Science* 169:487.
Levin, A.S., H.H. Fudenberg, J.E. Hopper, and S.K. Wilson (1971). "Immunofluorescent evidence for cellular control of synthesis of variable regions of light and heavy chains of immunoglobulins G and M by the same gene," *Proc. Nat. Acad. Sci. U.S.A.* 68:169.
Lichter, E.A. (1967). "Rabbit γA- and γM-immunoglobulins with allotypic specificities controlled by the *a* locus," *J. Immunol.* 98:139.
—— T.P. Conway, A. Gilman-Sachs, and S. Dray (1970). "Presence of allotypic specificities of three loci, *a*, *b*, and *f*, on individual molecules of rabbit colostral γA immunoglobulin," *J. Immunol.* 105:70.
Lieberman, R., and M. Potter (1969). "Crossing over between genes in the immunoglobulin heavy chain linkage group of the mouse," *J. Exp. Med.* 130:519.
Lobb, N., C.C. Curtain, and C. Kidson (1967). "Regulatory genes controlling the synthesis of heavy chains of human immunoglobulin-G," *Nature* 214:783.
Luzzati, A.L., R.M. Tosi, and A.O. Carbonara (1970). "Electrophoretically homogeneous antibody synthesized by spleen foci of irradiated repopulated mice," *J. Exp. Med.* 132:199.
MacDonald, A.B., and A. Nisonoff (1970). "Quantitative investigation of idiotypic antibodies. III. Persistence and variations of idiotypic specificities during the course of immunization," *J. Exp. Med.* 131:583.
Mach, B., H. Koblet, and D. Gros (1967). "Biosynthesis of immunoglobulin in a cell-free system," *Cold Spr. Hbr. Symp.* 32:269.
Mage, R.G. (1967). "Quantitative studies on the regulation of expression

of genes for immunoglobulin allotypes in heterozygous rabbits," *Cold Spr. Hbr. Symp.* 32:203.

—— and S. Dray (1965). "Persistent altered phenotypic expression of allelic γG-immunoglobulin allotypes in heterozygous rabbits exposed to isoantibodies in fetal and neonatal life," *J. Immunol.* 95:525.

—— —— (1966). "Persistence of altered expression of allelic G-immunoglobulin allotypes in an 'allotype suppressed' rabbit after immunization," *Nature* 212:699.

—— G.O. Young, and S. Dray (1967). "An effect upon the regulation of gene expression. Allotype suppression at the *a* locus in heterozygous offspring of immunized rabbits," *J. Immunol.* 98:502.

—— —— and R.A. Reisfeld (1968). "The association of the *c7* allotype of rabbits with some light polypeptide chains which lack *b* locus allotypy," *J. Immunol.* 101:617.

—— G.O. Young-Cooper, and C. Alexander (1971). "Genetic control of variable and constant regions of immunoglobulin heavy chains," *Nature* 230:63.

Makela, O. (1967). "The specificities of antibodies produced by single cells," *Cold Spr. Hbr. Symp.* 32:423.

—— (1970). "Analogies between lymphocyte receptors and the resulting humoral antibodies," *Transpl. Rev.* 5:3.

Mandy, W.J., and C.W. Todd (1969). "Characterization of allotype *A*11 in rabbits: a specificity detected by agglutination," *Immunochemistry* 6:811.

Mannik, M. (1967). "Variability in the specific interaction of H and L chains of γG-globulins," *Biochemistry* 6:134.

Marchalonis, J., and G.J.V. Nossal (1968). "Electrophoretic analysis of antibody produced by single cells," *Proc. Nat. Acad. Sci. U.S.A.* 61: 860.

Mason, S., and N.L. Warner (1970). "The immunoglobulin nature of the antigen recognition site on cells mediating transplantation immunity and delayed hypersensitivity," *J. Immunol.* 104:762.

Masuda, T., K. Kuribayashi, and M. Hanaoka (1969). "A new allotypic antigen of rabbit colostral A immunoglobulin," *J. Immunol.* 102:1156.

Matsuoka, Y., M. Takahashi, Y. Yagi, G.E. Moore, and D. Pressman (1968). "Synthesis and secretion of immunoglobulins by established cell lines of human hematopoietic origin," *J. Immunol.* 101:1111.

—— Y. Yagi, G.E. Moore, and D. Pressman (1969). "Differences in antigenicity of human IgG produced by three established lymphocytoid cell lines from a single individual," *J. Immunol.* 103:1176.

Merchant, B., and Z. Brahmi (1970). "Duplicate plating of immune cell products. Analysis of globulin class secreted by single cells," *Science* 167:69.

Mestecky, J., J. Zikan, and W.T. Butler (1971). "Immunoglobulin M and secretory immunoglobulin A: presence of a common polypeptide chain different from light chains," *Science* 171:1163.

Metzger, H. (1970). "Structure and function of γM macroglobulins," *Adv. Immunol.* 12:57.

Michael, J.G., and R. Marcus (1968). "Apparent production of two types of antibodies by a single cell," *Science* 159:1247.

Micheli, A., R.G. Mage, and R.A. Reisfeld (1968). "Direct demonstration and quantitation of *Aa*1, *Aa*2, and *Aa*3 allotypic specificities on Fd-fragments of rabbit immunoglobulin G," *J. Immunol.* 100:604.

Miller, J.F.A.P., and G.F. Mitchell (1968). "Cell to cell interactions in the immune response. I. Hemolysin-forming cells in neonatally thymectomized mice reconstituted with thymus or thoracic duct lymphocytes," *J. Exp. Med.* 128:801.

Miller, O.L., and B.R. Beatty (1969). "Visualization of nucleolar genes," *Science* 164:955.

Milstein, C. (1966a). "Chemical structure of light chains," *Proc. Royal Soc.*, ser. B, 166:138.

—— (1966b). "Variations in amino acid sequence near the disulfide bridges of Bence-Jones proteins," *Nature* 209:370.

—— (1966c). "Variations in amino acid sequence near the disulfide bridges of Bence-Jones proteins," *Biochem. J.* 101:338.

—— (1967). "Linked groups of residues in immunoglobulin kappa chains," *Nature* 216:330.

—— (1969a). "The basic sequences of immunoglobulin κ chains: sequence studies of Bence-Jones proteins Rad, Fr4, and B6," *Fed. Eur. Biochem. Soc. Letters* 2:301.

—— (1969b). "The variability of human immunoglobulin G," *Fed. Eur. Biochem. Soc. Symp.* 15:43.

—— B. Frangione, and J.R.L. Pink (1967). "Studies on the variability of immunoglobulin sequence," *Cold Spr. Hbr. Symp.* 32:31.

—— J.M. Jarvis, and C.P. Milstein (1969). "Occurrence of the Inv genetic markers in the subgroups of human kappa chains," *J. Mol. Biol.* 46:603.

—— C.P. Milstein, and A. Feinstein (1969). "Non-allelic nature of the basic sequences of normal immunoglobulin kappa chains," *Nature* 221:151.

—— and J.R.L. Pink (1970). "Structure and evolution of immunoglobulins," *Progr. Biophys. Mol. Biol.* 21:209.

Mitchell, G.F., and J.F.A.P. Miller (1968). "Cell to cell interaction in the immune response. II. Source of the hemolysin-forming cells in irradiated mice given bone marrow and thymus or thoracic duct lymphocytes," *J. Exp. Med.* 128:821.

Mitchison, N.A. (1967). "Antigen recognition responsible for induction in vitro of the secondary response," *Cold Spr. Hbr. Symp.* 32:431.

—— (1969). "Cellular and molecular recognition mechanisms prior to the immune response," in *Immunology*, suppl. to *Handbook of General Pathology* (New York, Springer-Verlag), chap. IV/A.

Mushinski, J.F., and M. Potter (1969). "Variations in leucine transfer ribonucleic acid in mouse plasma cell tumors producing kappa-type immunoglobulin light chains," *Biochemistry* 8:1684.

Natvig, J.B., H.G. Kunkel, and S.D. Litwin (1967). "Genetic markers of the heavy chain subgroups of human G globulin," *Cold Spr. Hbr. Symp.* 32:173.

Niall, H.D., and P. Edman (1967). "Two structurally distinct classes of kappa-chains in human immunoglobulins," *Nature* 216:262.

Nordin, A.A., H. Cosenza, and S. Sell (1970). "Immunoglobulin classes of antibody-forming cells in mice. II. Class restriction of plaque-forming cells demonstrated by replica-plating," *J. Immunol.* 104:495.

Nossal, G.J.V., A. Cunningham, G.F. Mitchell, and J.F.A.P. Miller (1968). "Cell to cell interactions in the immune response. III. Chromosome marker analysis of single antibody-forming cells in reconstituted, irradiated or thymectomized mice," *J. Exp. Med.* 128:839.

—— and J. Lederberg (1958). "Antibody production by single cells," *Nature* 181:1419.

Novotny, J., and F. Franek, (1970). "Tentative sequences of 52 amino acid residues from the constant part of pig immunoglobulin $\kappa$ chains," *Fed. Eur. Biochem. Soc. Letters* 9:33.

—— —— and F. Sorm (1970). "Large scale isolation, characterization and classification of pig immunoglobulin $\kappa$-chains," *Eur. J. Biochem.* 14:309.

Nussenzweig, V., I. Green, P. Vassali, and B. Benacerraf (1968). "Changes in the proportion of guinea-pig $\gamma_1$ and $\gamma_2$ antibodies during immunization and the cellular localization of these antigens," *Immunology* 14:601.

O'Donnell, I.J., B. Frangione, and R.R. Porter (1970). "The disulfide bonds of the heavy chain of rabbit immunoglobulin G," *Biochem. J.* 116:261.

Osoba, D. (1969). "Restriction of the capacity to respond to two antigens by single precursors of antibody producing cells in culture," *J. Exp. Med.* 129:141.

Oudin, J. (1960a). "Allotypy of rabbit serum proteins. I. Immunochemical analysis leading to the individualization of seven main allotypes," *J. Exp. Med.* 112:107.

—— (1960b). "Allotypy of rabbit serum proteins. II. Relationships between various allotypes: their common antigenic specificity, their distribution in a sample population; genetic implications," *J. Exp. Med.* 112:125.

—— (1966). "Genetic regulation of immunoglobulin synthesis," *J. Cell. Comp. Physiol.*, suppl. 1, 67:77.

—— and M. Michel (1969). "Idiotypy of rabbit antibodies. I. Comparison of idiotypy of antibodies against *Salmonella typhi* with that of antibodies against other bacteria in the same rabbits. II. Comparison of idiotypy of various kinds of antibodies formed in the same rabbits against *Salmonella typhi*," *J. Exp. Med.* 130:595 and 619.

Parkhouse, R.M.E., B.A. Askonas, and R.R. Dourmashkin (1970). "Electron microscopic studies of mouse immunoglobulin M; structure and reconstitution following reduction," *Immunology* 18:575.

Paul, C., A. Shimizu, H. Kohler, and F.W. Putnam (1971). "Structure of the hinge region of the mu heavy chain of human IgM immunoglobulins," *Science* 172:69.

Paul, W.E. (1970). "Functional specificity of antigen-binding receptors of lymphocytes," *Transpl. Rev.* 5:130.

Pauling, L. (1940). "A theory of the structure and process of formation of antibodies," *J. Am. Chem. Soc.* 62:2643.

Pernis, B., and G. Chiappino (1964). "Identification in human lymphoid tissues of cells that produce group 1 or group 2 gamma-globulins," *Immunology* 7:500.

—— —— A.S. Kelus, and P.G.H. Gell (1965). "Cellular localization of immunoglobulins with different allotypic specificities in rabbit lymphoid tissues," *J. Exp. Med.* 122:853.

—— L. Forni, and L. Amante (1970). "Immunoglobulin spots on the surface of rabbit lymphocytes," *J. Exp. Med.* 132:1001.

—— G. Torrigiani, L. Amante, A.S. Kelus, and J.J. Cebra (1969). "Identical allotypic markers of heavy polypeptide chains present in different immunoglobulin classes," *Immunology* 14:445.

Perry, M.B., and C. Milstein (1970). "Variability of interchain binding of immunoglobulins. Interchain bridge of human IgD," *Nature* 228:934.

Piggot, P.J., and E. M. Press (1967). "Cyanogen bromide cleavage and partial sequence of the heavy chain of a pathological human immunoglobulin," *Biochem. J.* 104:616.

Pink, J.R.L., S.H. Buttery, G.M. deVries, and C. Milstein (1970). "Human immunoglobulin subclasses. Partial amino acid sequence of the constant region of a $\gamma_4$ chain," *Biochem. J.* 117:33.

—— and C. Milstein (1967). "Inter heavy-light chain disulfide bridge in immune globulins," *Nature* 214:92.

—— —— (1969). "Sequence studies on a $\gamma_4$ immunoglobulin chain," *Fed. Eur. Biochem. Soc. Symp.* 15:177.

Pitcher, S.E., and W. Konigsberg (1970). "The sequence of the amino-terminal cyanogen bromide fragment from the heavy chain of a $\gamma G_1$ myeloma protein," *J. Biol. Chem.* 245:1267.

Ponstingl, H., and N. Hilschmann (1969). "Vollstaendig Aminosaeuresequenz einer λ-Kette der Subgruppe II (Bence-Jones Protein VIL). Die Evolution als Ursache der Antikoerper-Spezifitaet," *Z. Phys. Chem.* 350:1148.

—— J. Schwartz, W. Reichel, and N. Hilschmann (1970). "Die Primaerstruktur eines monoklonalen $\gamma_1$-Immunoglobulins (Myelomprotein Nie). I. Aminosaeuresequenz des variablen Teils der H-Kette, Subgruppen variabler Teile," *Z. Phys. Chem.* 351:1591.

Porter, R.R. (1959). "The hydrolysis of rabbit γ-globulin and antibodies with crystalline papain," *Biochem. J.* 73:119.

Potter, M., E. Appella, and S. Geisser (1965). "Variations in the heavy polypeptide chain structure of gamma myeloma immunoglobulins from an inbred strain of mice and a hypothesis as to their origin," *J. Mol. Biol.* 14:361.

—— W.J. Dreyer, E.L. Kuff, and K.R. McIntire (1964). "Heritable variation in Bence-Jones protein structure in an inbred strain of mouse," *J. Mol. Biol.* 8:814.

—— and R. Lieberman (1967). "Genetic studies of immunoglobulins in mice," *Cold Spr. Hbr. Symp.* 32:187.

Prahl, J.W. (1967). "The C-terminal sequences of the heavy chains of human immunoglobulin G myeloma proteins of differing isotypes and allotypes," *Biochem. J.* 105:1019.

—— W.J. Mandy, G.S. David, M.W. Steward, and C.W. Todd (1970). "Participation of allotypic markers in rabbit immunoglobulin classes," *Protides of the Biol. Fluids* 17:125.

—— —— and C.W. Todd (1969). "The molecular determinants of the *A* 11 and *A* 12 allotypic specificities in rabbit immunoglobulin," *Biochemistry* 8:4935.

—— and R.R. Porter (1968). "Allotype-related sequence variation of the heavy chain of rabbit immunoglobulin G," *Biochem. J.* 107:753.

Press, E.M., and N.M. Hogg (1969). "Comparative study of two immunoglobulin G Fd fragments," *Nature* 223:807.

—— —— (1970). "The amino acid sequence of the Fd fragments of two human $\gamma$ heavy chains," *Biochem. J.* 117:641.

Quattrocchi, R., D. Cioli, and C. Baglioni (1969). "A study of immunoglobulin structure. III. An estimation of the variability of human light chain," *J. Exp. Med.* 130:401.

Raff, M.C. (1970). "Role of thymus derived lymphocytes in the secondary humoral immune response in mice," *Nature* 226:1257.

Rajewsky, K., V. Schirrmacher, S. Nase, and N.K. Jerne (1969). "The requirement of more than one antigenic determinant for immunogenicity," *J. Exp. Med.* 129:1131.

Reisfeld, R.A. (1967). "Heterogeneity of rabbit light-polypeptide chains," *Cold Spr. Hbr. Symp.* 32:291.

—— S. Dray, and A. Nisonoff (1965). "Differences in amino acid composition of rabbit G-immunoglobulin light polypeptide chains controlled by allelic genes," *Immunochemistry* 2:155.

—— and J.K. Inman (1968). "Amino acid composition of rabbit IgG light chains with *b*6 allotypic specificity," *Immunochemistry* 5:503.

Rowe, D.S., F. Dolder, and H.D. Welscher (1969). "Studies on human IgD. I. Molecular weight and sedimentation coefficient," *Immunochemistry* 6:437.

Ruffili, A., A. Compere, and C. Baglioni (1970). "Repression of the synthesis of immunoglobulins in newborn mice by antibodies directed against the variable region of light chains," *J. Immunol.* 105:1511.

Sager, R., and F.J. Ryan (1961). *Cell Heredity* (New York, Wiley), Table 6.4.

Saha, A., P. Chowdhury, S. Sambury, Y. Behelak, D.C. Heiner, and B. Rose (1970). "Studies on human IgD. II. Physicochemical characterization of human IgD," *J. Immunol.* 105:238.

Sayers, D.L. (1923). *Whose Body?* (New York, Harper), pp. 96–97.

Schalet, A. (1969). "Exchanges at the bobbed locus of *Drosophila Melanogaster,*" *Genetics* 63:133.

Schiechl, H., and N. Hilschmann (1971). "Die Primaerstruktur einer monoklonalen Immunoglobulin-L-Kette der Subgruppe I vom K-Typ (Bence-Jones-Protein Au): gekoppelte Austausche innerhalb der Subgruppen," *Z. Phys. Chem.* 352:111.

Schimpl, A., and E. Wecker (1970). "Inhibition of *in vitro* immune response by treatment of spleen cell suspensions with anti $\theta$ serum," *Nature* 226:1258.

Schlesinger, M. (1970). "Anti-$\theta$ antibodies for detecting thymus dependent lymphocytes in the immune response of mice to SRBC," *Nature* 226:1254.

Schubert, D., and M. Cohn (1970). "Immunoglobulin biosynthesis. V. Light chain assembly," *J. Mol. Biol.* 53:305.

Segre, M., D. Segre, and F.P. Inman (1969). "Comparison of A*a*1 allotypic specificity carried by rabbit IgG and IgM," *J. Immunol.* 102:1368.

Sell, S. (1966). "Immunoglobulin M allotypes of the rabbit. Identification of a second specificity," *Science* 153:641.

—— (1967). "Isolation and characterization of rabbit colostral IgA," *Immunochemistry* 4:49.

—— (1970). "Development of restrictions in the expression of immunoglobulin specificities by lymphoid cells," *Transpl. Rev.* 5:19.

—— and S.J. Hughes (1968). "Further characterization of rabbit immunoglobulin allotype *Ab*9," *Immunochemistry* 5:401.

Shearer, G.M., and G. Cudkowicz (1969). "Distinct events in the immune response elicited by transferred marrow and thymus cells. I. Antigen requirements and proliferation of thymic antigen-reactive cells," *J. Exp. Med.* 130:1243.

—— —— and R.L. Priore (1969). "Cellular differentiation of the immune system of mice. IV. Lack of class differentiation in thymic antigen-reactive cells," *J. Exp. Med.* 130:467.

Shelton, E., and K.R. McIntire (1970). "Ultrastructure of the $\gamma$M immunoglobulin molecule," *J. Mol. Biol.* 47:595.

Shimizu, A., C. Paul, H. Kohler, T. Shinada, and F.W. Putnam (1971). "Variation and homology in the mu and gamma heavy chains of human immunoglobulins," *Science* 173:629.

Simonsen, M. (1967). "The clonal selection hypothesis evaluated by grafted cells reacting against their hosts," *Cold Spr. Hbr. Symp.* 32:517.

Siskind, G.W., and B. Benacerraf (1969). "Cell selection by antigen in the immune response," *Adv. Immunol.* 10:1.

Small, P.A., R.A. Reisfeld, and S. Dray (1965). "Peptide differences of rabbit $\gamma$G immunoglobulin light chains controlled by allelic genes," *J. Mol. Biol.* 11:713.

Smith, G.P., L. Hood, and W.M. Fitch (1971). "Antibody diversity," *Ann. Rev. Biochem.* 40:969.

Smithies, O. (1963). "Gamma-globulin variability: a genetic hypothesis," *Nature* 199:1231.
—— (1967a). "The genetic basis of antibody variability," *Cold Spr. Hbr. Symp.* 32:161.
—— (1967b). "Antibody variability," *Science* 157:267.
—— (1970). "Pathways through networks of branched DNA," *Science* 169:882.
—— D.M. Gibson, and E.M. Fanning (1972). Reported in *Atlas of Protein Sequence and Structure,* vol. 5, M.O. Dayhoff, ed. (Silver Spring, National Biomedical Research Foundation).
—— —— —— M.E. Percy, D.M. Parr, and G.E. Connell (1970). "Deletions in immunoglobulin polypeptide chains as evidence for breakage and repair in DNA," *Science* 172:574.
—— —— and M. Levanon (1970). "Linkage relationships in normal light chains," in *Developmental Aspects of Antibody Formation and Structure* (Prague, Academia), p. 339.
Spiegelberg, H.L., J.W. Prahl, and H.M. Grey (1970). "Structural studies of human γD myeloma protein," *Biochemistry* 9:2115.
Steinberg, A.G., W.A. Muir, and S.A. McIntire (1968). "Two unusual *GM* alleles. Their implications for the genetics of the *Gm* antigens," *Am. J. Human Genetics* 20:258.
Steiner, L.A., and R.R. Porter (1967). "The interchain disulfide bonds of a human pathological immunoglobulin," *Biochemistry* 6:3957.
Stemke, G.W. (1964). "Allotypic specificities of A- and B-chains of rabbit gamma globulin," *Science* 145:403.
—— and R.J. Fischer (1965). "Rabbit 19s antibodies with allotypic specificities of the *a*-locus group," *Science* 150:1298.
Strehler, B.J., D.D. Hendley, and G.P. Hirsch (1967). "Evidence on a codon restriction hypothesis of cellular differentiation: multiplicity of mammalian leucyl-sRNA-specific synthetases and tissue-specific deficiency in an alanyl-sRNA synthetase," *Proc. Nat. Acad. Sci. U.S.A.* 57:1751.
Svasti, J., and C. Milstein (1970). "Variability of interchain binding of immunoglobulins. Interchain bridges of mouse $IgG_1$," *Nature* 228:932.
Svehag, S.E., and B. Bloth (1970). "Ultrastructure of secretory and high-polymer serum immunoglobulin A of horse and pig origin," *Science* 168:847.
—— —— and M. Seligmann (1969). "Ultrastructure of papain and pepsin digestion fragments of human IgM globulins," *J. Exp. Med.* 130:691.
—— B. Chesebro, and B. Bloth (1967). "Ultrastructure of gamma M immunoglobulin and alpha macroglobulin: electron microscopic study," *Science* 158:933.
Takahashi, M., N. Takagi, Y. Yagi, G.E. Moore, and D. Pressman (1969a). "Immunoglobulin production in cloned sublines of a human lymphocytoid cell line," *J. Immunol.* 102:1388.
—— N. Tanigaki, Y. Yagi, G.E. Moore, and D. Pressman (1968). "Presence

of two different immunoglobulin heavy chains in individual cells of established human hematopoietic cell lines," *J. Immunol.* 100:1176.

—— Y. Yagi, G.E. Moore, and D. Pressman (1969b). "Pattern of immunoglobulin production in individual cells of human hematopoietic origin in established culture," *J. Immunol.* 102:1274.

Talmage, D.W. (1959). "Immunological specificity," *Science* 129:1643.

Terry, W.D., L. Hood, and A.G. Steinberg (1969). "Genetics of immunoglobulin κ-chains: chemical analysis of normal human light chains of differing Inv types," *Proc. Nat. Acad. Sci. U.S.A.* 63:71.

—— B.W. Matthews, and D.R. Davies (1968). "Crystallographic studies of a human immunoglobulin," *Nature* 220:239.

—— and J.J. Ohms (1970). "Implications of heavy chain disease protein sequences for multiple gene theories of immunoglobulin synthesis," *Proc. Nat. Acad. Sci. U.S.A.* 66:558.

Thorpe, N.O., and H.F. Deutsch (1969). "Studies on papain produced subunits of human γG-globulin. II. Structure of peptides related to the genetic *Gm* activity of γG-globulin Fc-fragments," *Immunochemistry* 3:329.

Titani, K., T. Shinoda, and F.W. Putnam (1969). "The amino acid sequence of a κ type Bence-Jones protein. III. The complete sequence and the location of the disulfide bridges," *J. Biol. Chem.* 244:3550.

Todd, C.W. (1963). "Allotypy in rabbit 19s protein," *Biochem. Biophys. Res. Comm.* 11:170.

—— and F.P. Inman (1967). "Comparison of the allotypic combining sites on H-chains of rabbit IgG and IgM," *Immunochemistry* 4:407.

Tomasi, T.B., and J. Beinenstock (1968). "Secretory immunoglobulins," *Adv. Immunol.* 9:1.

Tosi, S.L., R.G. Mage, and S. Dubiski (1970). "Distribution of allotypic specificities *a*1, *a*2, *a*14, and *a*15 among immunoglobulin G molecules," *J. Immunol.* 104:641.

Trump, G.N. (1970). "Goldfish immunoglobulins and antibodies to bovine serum albumin," *J. Immunol.* 104:1267.

Turner, M.W., and H. Bennich (1968). "Subfragments from the Fc fragment of human immunoglobulin G," *Biochem. J.* 107:171.

—— L. Martensson, J.B. Natvig, and H. Bennich (1969). "Genetic (*Gm*) antigens associated with subfragments from the Fc fragment of human immunoglobulin G," *Nature* 221:1166.

Utsumi, S. (1969). "Stepwise cleavage of rabbit immunoglobulin G by papain and isolation of 4 types of biologically active Fc fragments," *Biochem. J.* 112:343.

Valentine, R.C., and N.M. Green (1967). "Electron microscopy of an antibody-hapten complex," *J. Mol. Biol.* 27:615.

Vice, J.L., A. Gilman–Sachs, W.L. Hunt, and S. Dray (1970). "Allotype suppression in *a*2*a*2 homozygous rabbits fostered *in uteri* of *a*2-immunized *a*1*a*1 homozygous mother and injected at birth with anti *a*2 antiserum," *J. Immunol.* 104:550.

—— W.L. Hunt, and S. Dray (1969). "Allotype suppression with anti-*b*5 antiserum in *b*5*b*5 homozygous rabbits fostered *in uteri* of *b*4*b*4 homo-

zygous mothers: compensation by allotypes of other loci," *J. Immunol.* 103:629.
—— —— —— (1970). "Contribution of the *b* and *c* light chain loci to the composition of rabbit γG immunoglobulins," *J. Immunol.* 104:38.
Wallace, H., and M. Birnsteil (1966). "Ribosomal cistrons and the nucleolar organizer," *Biochim. Biophys. Acta* 114:296.
Wang, A.C., and H.H. Fudenberg (1969). "Genetic control of gamma chain synthesis: a chemical and evolutionary study of the *Gm(a)* factor of immunoglobulins," *J. Mol. Biol.* 44:493.
—— J.R.L. Pink, H.H. Fudenberg, and J.J. Ohms (1970a). "A variable region subclass of heavy chain common to immunoglobulins G, A, and M and characterized by an unblocked amino-terminal residue," *Proc. Nat. Acad. Sci. U.S.A.* 66:657.
—— I. Y. F. Wang, J.N. McCormick, and H.H. Fudenberg (1969). "The identity of the light chains of monoclonal IgG and IgM in one patient," *Immunochemistry* 6:451.
—— S.K. Wilson, J. E. Hopper, H.H. Fudenberg, and A. Nisonoff (1970b). "Evidence for control of synthesis of the variable regions of the heavy chains of immunoglobulins G and M by the same gene," *Proc. Nat. Acad. Sci. U.S.A.* 66:337.
Warner, N.L., P. Byrt, and G.L. Ada (1970). "Blocking of the lymphocyte antigen receptor site with anti-immunoglobulin sera in vitro," *Nature* 226:942.
—— L.A. Herzenberg, and G. Goldstein (1966). "Immunoglobulin isoantigens (allotypes) in the mouse. II. Allotype analysis of three $\gamma G_2$-myeloma proteins from (NZB x BALB/c) F1 hybrids and of normal $\gamma G_2$-globulins," *J. Exp. Med.* 123:707.
Watanabe, S., and N. Hilschmann (1970). "Die Primaerstruktur einer monoklonalen Immunoglobulin-L-Kette der Subgruppe I vom κ-Typ (Bence-Jones-Protein Hau): Untergruppen innerhalb der Subgruppen," *Z. Phys. Chem.* 351:1291.
Watson, J.D. (1970). *Molecular Biology of the Gene,* 2nd ed. (New York, Benjamin).
Weigert, M.G., I.M. Cesari, S.J. Yonkovich, and M. Cohn (1970). "Variability in the lambda light chain sequences of mouse antibody," *Nature* 228:1045.
Weiler, E. (1965). "Differential activity of allelic γ-globulin genes in antibody-producing cells," *Proc. Nat. Acad. Sci. U.S.A.* 54:1765.
—— and I.J. Weiler (1968). "Unequal association of mouse allotypes with antibodies of different specificities," *J. Immunol.* 101:1044.
Wensink, P.C., and D.D. Brown (1971). "Denaturation map of the ribosomal DNA of *Xenopus laevis*," *J. Mol. Biol.* 60:235.
Whitney, P.L., and C. Tanford (1965). "Recovery of specific activity after complete unfolding and reduction of an antibody fragment," *Proc. Nat. Acad. Sci. U.S.A.* 53:524.
Wigzell, H. (1970). "Specific fractionation of immunocompetent cells," *Transpl. Rev.* 5:76.
Wikler, M., K. Titani, T. Shinoda, and F.W. Putnam (1967). "The com-

plete amino acid sequence of a λ type Bence-Jones protein," *J. Biol. Chem.* 242:1668.
Wilheim, E., and M.E. Lamm (1966). "Absence of allotype *b*4 in the heavy chain of rabbit γG-immunoglobulin," *Nature* 212:846.
Wilkinson, J.M. (1969a). "Variation in the N-terminal sequence of heavy chains of immunoglobulin G from rabbits of different allotype," *Biochem. J.* 112:173.
—— (1969b). "α-chain of immunoglobulin A from rabbits of different allotype: composition and N-terminal sequence," *Nature* 223:616.
—— (1970). "Genetic markers of rabbit immunoglobulins," *Biochem. J.* 117:3P.
Williamson, R. (1970), at the Eighth International Congress of Biochemists, Montreux. Quoted by Brown, Wensink, and Jordan (1971).
Wimber, D.E., and D.M. Steffensen (1970). "Localization of 5*S* RNA genes on *Drosophila* chromosomes by RNA-DNA hybridization," *Science* 170:639.
Wu, T.T., and E.A. Kabat (1970). "An analysis of the sequences of the variable regions of Bence-Jones proteins and myeloma light chains and their implications for antibody complementarity," *J. Exp. Med.* 132:211.
Yang, W.K., and G.D. Novelli (1968). "Studies on the multiple iso-accepting transfer ribonucleic acids in mouse plasma cell tumors," in *Nucleic Acids in Immunology* (New York, Springer-Verlag), p. 644.
Zoschke, D.C., and F.H. Bach (1970). "Specificity of antigen recognition by human lymphocytes *in vitro*," *Science* 170:1404.
—— —— (1971). "Specificity of allogeneic cell recognition by human lymphocytes *in vitro*," *Science* 172:1350.
Zullo, D.M., C.W. Todd, and W.J. Mandy (1968). *Proc. Canad. Fed. Biol. Soc. Abstr.* 11:111. Quoted by Mandy and Todd (1969).
Zvaifler, N.J., and J.O. Robinson (1969). "Rabbit homocytotropic antibody. A unique rabbit immunoglobulin analogous to human IgE," *J. Immunol.* 130:907.

# Index

| DATE | | | |
| --- | --- | --- | --- |
| | | | |
| | | | |
| | | | |
| | | | |
| | | | |
| | | | |
| | | | |
| | | | |
| | | | |
| | | | |
| | | | |
| | | | |